American Institute of Chemical Engineers

Management Division

Presentations at the
2007 AIChE Annual Meeting

November 4-9, 2007
Salt Lake City, Utah, USA

Printed from e-media with permission by:

Curran Associates, Inc.
57 Morehouse Lane
Red Hook, NY 12571
www.proceedings.com

ISBN: 978-1-60560-003-1

Some format issues inherent in the e-media version may also appear in this print version.

ISBN: 978-1-60560-003-1

American Institute of Chemical Engineers
Proceedings
Three Park Avenue
New York, NY 10016-5991
Phone: 212-591-8100

www.aiche.org

American Institute of Chemical Engineers

Management Division

Presentations at the
2007 AIChE Annual Meeting

AIChE

November 4-9, 2007
Salt Lake City, Utah, USA

American Institute of Chemical Engineers

Management Division

TABLE OF CONTENTS

167a Patent Strategies to Generate/Manage Disprutive Innovation

Neal Seth, Baker Hostetler LLP, Washingon Square, Suite 1100, 1050 Connecticut Ave. NW, Washington, DC 20036

This talk will analyze patent filing and enforcement strategies to generate/manage disruptive technologies.

WORK PROCESSES AND PEOPLE EXCELLENCE: THE LAST COMPETITIVE ADVANTAGES IN THE CHEMICAL INDUSTRY

Delia Contreras and Fabio Bravo
The Dow Chemical Company
E-mails: Contreras@Dow.com
FBravo@Dow.com

At the current stage of the globalization process, it has become clear that competitive advantages like technology, access to raw materials and capital, which used to be important competitive advantages in the chemical industry, have become less and less significant. Technology can be easily purchased or developed, raw materials are now available to almost anybody and there is a relatively easy flow of capital from one place to another. Therefore, for chemical and petrochemical companies, there are only two remaining competitive advantages: work processes and people excellence. This paper will elaborate on these points and will discuss what companies can do to maximize their competitive advantage in these two areas.

People as Competitive Advantage

For years we have heard that the greatest asset of any company is their people. However, sometimes and in certain industries / companies, it has probably been more of a nice saying than a reality (or at least that was the perception from some employees). However, now it is for real because it is not anymore a matter of choice: there is no other option for companies than to fully understand this and to engrain it in their culture. It is probably not much change after all. The real difference is that globalization has made people much more important and visible due to realities like international fierce competition, people's mobility, open flow of ideas and capital, and shortage of qualified people.

Regardless of size, capital or industry, if a company cannot attract and retain the right talent it will become unable to compete and will eventually fail (1). People's knowledge, skills, experience, brain-power and potential are the most important assets of any company. Clearly, once you have those you can generate profits, attract investors, select or develop technologies. The value of a good company is much higher than just the book value of the company's assets; what is the difference? The know-how, good-will, knowledge and capabilities of its people; bottom-line: people's brains.

Once we accept those realities, it is easy to conclude that human resources (HR) departments are key functions in any organization; HR excellence (which includes honesty, flexibility and openness) is a must for companies to attract, retain, develop and motivate the best people in each market. Therefore, nowadays no company can afford to have a Dilbert's world type HR function, one that instead of being a competitive advantage pretty much becomes a burden and an employee irritating nuisance. One way to achieve this is to assign great people with background in other functions to HR and provide them with opportunities for advancement instead of making HR the road's end.

We all realize that globalization has brought a fierce competition to the chemical industry and in general to the business world. What sometimes we have failed to realize is that competition for talent is even fiercer. Some people cannot understand this considering that globalization brought to the table literally billions of people from China, India, Russia, Latin America. Definitely there is no shortage of people … there is shortage of talent, shortage of qualified people who can perform a high level in an extremely competitive environment. Reasons for that shortage of talent include globalization itself, as there are now tremendous needs for qualified and experienced people all over the world. It has been estimated that China will require 75,000 top-level executives by 2010, about 70,000 more than they have now (2). Another reason for shortage of talent is the baby-boomer generation starting to massively retire in mature markets like US (with about 76 million baby-boomers) and Europe. In fact, "the country's 500 biggest companies anticipate losing half their senior management in the next five to six years" (3). This "exodus" in the most developed economies results not only in reduction of available qualified workforce but also makes it much more diverse than it is today.

So, the ability to find, fully develop and fully utilize talented people plus ability to retain such people is the key not only for growth but pretty much for survival. But, how do you retain people? Does it mean the companies need to relax the standards so that people stay? Not really … there are some minimum requirements for companies to be able to retain the best talent such as competitive salary and benefits package, a culture of respect for people, high ethical standards. Clearly without those any retention efforts will fail. Giving people some flexibility in working hours and location will help with retention. However, the "high power" retention tools are providing meaningful and challenging jobs, having clear career paths, creating an environment in which people can perform at their fullest capacity (not just to the level management expects them to perform) and having a performance evaluation system that properly compares each worker against his/ her colleagues and compensate them accordingly. Bear in mind that people want and need to be measured … competitiveness seems to be in human nature. Although in some cultures the emphasis is in individual performance and others in team performance, competitiveness and desire for measurement (as team or as individual) seems universal. The metrics must be fair, clear, applied to everybody and meaningful regardless of the geographic location. Creating those "global" metrics can be challenging but, again, absolutely necessary.

As a corollary of the paragraphs above, when a person quits his / her job to go to another company, it should be taken very seriously as literally would be a big loss in terms of money. Additionally, "you never know, when somebody leaves, if she would have been the next rainmaker" (4). Therefore, as in any serious loss, a careful investigation should be conducted to discover the root cause of this loss and appropriate measures should be taken to avoid for this kind of loss to happen again.

To summarize this section, it is quite clear that in today's environment people's excellence is one of the two remaining sustainable competitive advantages. This is because of two basic reasons: first because the other "competitive advantages" are pretty much not sustainable (i.e., competitors can develop or purchase technology, capital can easily flow from country to country or company to company). Second, because your people are the ones that will come up with the next idea and will develop the next best technology so they are really a continuous source of competitive advantage (or disadvantage if the conditions are such).

Work Processes as Competitive Advantage

Understanding work processes as the only other competitive advantage is sometimes more difficult. Let's start by pointing-out that if you have good people but not effective work-processes, you may not get sustainable results. Yes, people can change work processes, but normally it takes long time to do so and by the time it gets done your best people may have lost motivation or got out of track or may have even moved out of the department or company due to frustration. In summary, you need effective work processes to allow people to perform at their fullest capacity. Work processes can greatly improve productivity and are the means for companies to consistently achieve positive results. However, work processes need to be flexible enough to allow for innovation and improvement and to, in that way, drive toward excellence. In fact, work processes cannot be "straight jackets" and need to leave some room for people to maneuver depending on the circumstances; they also need to be streamlined so that they don't add bureaucracy. Work process not only provide structure and discipline to help companies to improve performance but will also help companies to deal with crisis in a more effective manner.

We are talking here about work processes in general, not just manufacturing work processes, as work processes in areas like engineering, supply chain, marketing analysis, hiring and promoting people, sharing information, billing the customer can all be significant competitive advantages. In general terms, any company's activity that is prone to significant variance (and most activities involving human beings are) becomes a work process implementation opportunity as a mechanism to reduce variance and to standardize in the most advantageous practices.

Another reason for which work processes have become a key competitive advantage is globalization, because globalization has accentuated the need for consistent products, as global customers need products with exactly the same characteristics in each market. In other words, the pound of polyethylene that you sell to a global customer in Colombia needs to be pretty much identical to the one that you sell to the same global customer in the US, Germany or China. For other industries (consumer products, food processing, automobile) that is not always the case as you need to consider local tastes and preferences; for the chemical industry the products tend to be more "universal" and therefore "customization" is less relevant. However, somehow you need to adjust to local culture and local regulations so you need some degree of flexibility or customization, making sure that every plant, regardless of location, meets the intent but not necessarily follows all the work process in exactly the same way. Another way is to have work processes that take into account the differences (means getting input from everybody in advance) … problem is that you can get into the "lowest denominator" syndrome and have work processes that are not as effective as needed. Achieving consistency is a serious challenge for any company in any industry, however achieving consistency by a multi-national chemical company with operations all around the world is really a gigantic challenge (5).

So, how do you get consistent chemical products? The first steps are consistency of raw materials and equipment / technology …all challenging but achievable. The most difficult part, however, is to get consistency of behavior during manufacturing (5). This consistency is much more difficult to achieve because deals with cultural differences and local practices.

What about Other Competitive Advantages?

What about more "traditional" competitive advantages like "customer captivity" and economy of scale? Well, we are not going to get into details on those but let's explore a little bit. We all know that customer captivity is a very perishable competitive advantage in any industry and, in any case, a very weak one in the chemical industry. In fact, by definition, customer loyalty is almost nonexistent in the commodity chemicals. Besides, it is not that significant in the specialty chemicals, unless you have a new or improved product that nobody has matched yet. So, this means that in the chemical industry a company has to prove itself every day with the customers … in fact, the ability to do so is the competitive advantage. And how does a company prove itself every day? By having the best people and best work processes so that it can deliver high quality material, on time and at the lowest possible price. So, bottom-line the competitive advantage again is having the best people and the best work processes in all functions to satisfy and exceed customer needs in terms of quality, timing and price.

Regarding economy of scale, let's start by saying that economy of scale can be a huge competitive advantage when properly combined with other factors but it can be argued that "enormous size alone does not deliver a competitive advantage" (6). The classical example of economy of scale as competitive advantage is Wal-Mart due to its well known ability to negotiate large discounts due to its enormous size. However, reference (6) explains how Wal-Mart can be even better counter-example when operating margins (defined as earnings before interest and taxes) are analyzed. Wal-Mart's operating margins have decreased continuously from the 7-8% of sales in the mid 80's when Wal-Mart was a much smaller company to about 5% nowadays, including international operations and Sam's clubs. You would expect that the spectacular growth of Wal-Mart in those 20+ years would result in much better leverage to negotiate big discounts and in that way significantly increase the margins, but history shows that it has not happened. Another way to illustrate that economy of scale by itself is not a competitive advantage is to realize how many successful small companies struggle and fail when they start growing. Economy of scale, however, becomes a huge competitive advantage in the chemical industry when it translates into better ability to serve the customer in terms of quality, timing and price. In other words, size becomes a tremendous competitive advantage when it helps to better serve the customer, which can be achieved via vertical integration, ability to develop or purchase the best technologies, capability to attract, retain and develop the best people, ability to compete globally, ability to maintain market share in a changing market, capability to serve global customers, ability to develop, implement and improve the best work processes, ability to participate in multiple joint-ventures, better access to capital to grow and enter new markets, etc. Of course, we all know that the chemical industry implies large facilities and huge investments as we also know that it is very expensive to produce a pound of polyethylene in the lab. So, certain minimum size, which would be considered huge size in other fields, is required just to be a small player in the chemical industry. However, after certain size, economy of scale by itself is at the most a weak competitive advantage but could become a tremendous competitive advantage if it means better capability to serve the customer (in terms of quality, timing and price), access to the best talent pool, capability to develop and implement the best work process, etc.

Diversity and Consistency

We can easily explain why consistency is important in the chemical industry. To explain why diversity is important probably is sufficient to say that companies need diversity of thinking to have different perspectives and generate enough ideas to effectively solve problems and continuously improve/innovate in all areas. Teams in which all members have similar backgrounds and opinions can easily miss the whole picture and/or miss opportunities. But, can you achieve both diversity and consistency? If not, which one is more important? We will go back to the former question later on but let's try to answer the latter one. So, which one is more important, diversity or consistency? The answer is "it depends". For example, consistency is much more critical when you are starting operations at a new location or while trying to integrate a recently acquired operation or company. Reason is because at that point the key is to get the people in the right and safe mindset, to allow them to quickly learn the processes and procedures required to produce the consistent goods. After some time, when you have reached what we could call "steady-state", the focus should change to improvement and innovation. There are two logical reasons for such approach: first, plain good timing, as at such point consistency has been "achieved" and the right behaviors are already "engrained in the culture". The second and most important reason is that if any business remains static in today's fierce competitive environment, sooner or later would result in getting behind and loosing market-share. In fact, the only constant in today's business world is change and companies and people that do not feel comfortable with change will definitely struggle.

The relative importance of diversity vs. consistency also depends on function. For example, you need consistency of operations to get consistent products; however enforcing consistency in R&D would probably lead to mediocre results (to say the least). The real key to success for a company is to enforce consistency in some functions / activities and to encourage diversity in some others.

We hear a lot about cultural differences, cultural shock, etc. partially because we tend to notice the differences very early when we get exposed to another culture, but as human beings probably we are much more alike than we are different as we have pretty much the same common human desires and needs. For example we all have the same basic needs (food, clothing, housing, sense of security, etc) and pretty much the same drivers (looking for meaning in life, need for social acceptance, desire to connect with others, etc). Leaders need to properly understand what is pretty much universal to all human beings and what is unique to each culture. Superficial understanding of other cultures can have tremendous negative impacts on businesses, for that reason each company needs local brainpower plus some flexibility to adapt to local environments (7). On the other hand, leaders and "company cultures" needs to "borrow" the best of each culture (and try to disregard the worst of each culture). The real "diversity and consistency" commitment of a company is its ability to leverage diverse experiences, backgrounds, problem-solving and innovation skills, so that everybody, regardless of background, can contribute to his fullest capacity and will receive recognition showing that his/her efforts are worthwhile and appreciated.

Companies need to leave some freedom for locals to incorporate their own cultural needs, especially considering that the most localized a strategy is, the most chances that effective execution is achieved. What are the boundaries for customizing? There are three types of boundaries: a) when local practices are not delivering consistent products or somehow do not meet the intend, b) when local practices deviate from corporate vision or goals, c) when local practices do not follow ethical standards defined by the company.

Improved Consistency and Innovation

At first look the phrase "improved consistency" may look like an oxymoron but it is not. The primary goal is to achieve consistency which is probably not that difficult to get when there are mechanisms in place to enforce that the proper work processes are followed and consistency of behavior is achieved. Each unit and each geographic location must help develop, use and share the best practices. The real challenge is to make sure that the work processes allow and promote improvement and innovation and that whatever improvement or innovation is proved at a given facility/location, gets leveraged at other facilities/locations. Then the already improved work process will receive further improvement and will be leveraged, again and again, across the company … that is the real path to operational excellence and what we call here improved consistency.

Companies need to facilitate "global conversations" to share best practices and to build-up on them to constantly improve. Key word is conversation, meaning two way exchange, not one party "imposing" without listening. So, the basis for this conversation is the understanding that different people bring their own experiences, knowledge, set of values, capabilities … and it is critical to leverage on those to really achieve improvement and innovation. Another key is to understand that nobody by itself is as smart (never mind smarter) as the entire group; so it makes no sense for somebody to try to impose his/her points of view. Leaders and workers in general, need to become good listeners, respect others' opinions and try hard to address and incorporate those opinions. No company wants or can afford to lose those good ideas because a leader is not a good listener or because he/she underestimates the person that came up with a good idea.

The strong competition in the chemical industry is a clear call for "innovate to grow" and even "innovate to survive". Nowadays, no company can remain stagnant without risking its market share. Besides, considering the competition and low margins that globalization have brought to most industries, there is no other option than to innovate. So, the real question is not "to innovate or not to innovate?" but "innovate in what?" Are we just talking about new products and new technologies? Well, development of new products and new technologies is a must, but it is not all what is needed. Besides, in the chemical industry new products and new technologies tend to be an expensive and long-term proposition and therefore cannot occur as frequently as in other industries. Companies need to innovate in every area, including work processes; attraction and retention; cost reduction; interaction with clients; how to do business; *how to get the best of everybody*; how to continuously improve, how to find the best partners for joint-ventures, etc.. Innovation in these areas is typically less capital intensive, faster and less dramatic than developing new products or new technologies, but offers plenty of opportunities to contribute to the company's bottom-line and to the company's growth.

Let's talk a little bit about the phrase "get *the best of everybody*", what does it really mean? It means companies need to allow and encourage everybody in the organization to become problem solvers, improvers and innovators. This is not common in some US companies where, for example, the "first line" employees are not always expected to use their full brain power; that is a big mistake! First, because they are the closest people to the operation and therefore they have the best knowledge of what exactly the problems are and could quickly act on them before the problems get more serious. Second reason is that with the strong competition that we are facing these days, no company can afford not to engage everybody's full capabilities ... especially when we know that some competitors have strong programs to assure that they do so. An example from the automobile industry is Toyota's ability to fully engage the intellect of all employees in their improvement process as Toyota gives them skills, tools, and permission to solve the problems in real-time and also to prevent problems before they occur. Some American automobile companies rely more in the staff or the so called "experts" to solve problems and to make improvements ... needless to say whose strategy has been more successful.

Innovation has become the "buzz word". It is pretty much accepted that nowadays innovation is a must for any company to grow and even to survive. The billion-dollar question is how to get a company into the innovation mode? Some companies have created high level "Chief Innovation Officer" positions (8) to boost their innovation capabilities. Some companies (and even countries) have engaged innovation "gurus" to help with that (9). Some companies have recognized the importance of the customer in the innovation process as they need to fully understand what the customer really wants (10); therefore voice-of-the-customer exercises have become quite common in the innovation process. Other ways are "engaging" everybody's brains, encouraging and allowing people to "think out of the box", promoting diversity (especially diversity of thinking) and inclusion, leaving some room for experimentation and "failure" (remember that if there are no occasional "failures" you may not be pushing hard enough), adding innovativeness to the performance appraisal (11), creating a culture that not only promotes idea generation but that quickly acts on those ideas … nothing kills innovation faster than lack of response or action once the ideas are formulated.

Summary

In summary, in the chemical industry (as in pretty much all industries), people excellence has become a necessary condition not only for growth but also for sustainability. Considering the limited talent pool, companies must become excellent at attracting and retaining the best people in every discipline and geographical area. Such talent, enabled by constantly improved work processes, will drive the company to excellence at all levels and innovation, which are absolutely necessary for sustainability and growth.

DISCLAIMER: the ideas expressed in this paper correspond to authors' understanding and do not necessarily match the vision and corporate culture of The Dow Chemical Company.

REFERENCES

(1) ULRICH, Dave and Dale Lake, "Organizational Capability", John Wiley & sons, New York, 1990.
(2) LAELLE, Louis, "China's B-School Boom", Business Week, January 9/2006, pp. 40-43.
(3) BYRNES, Nanette, "Star Search", Business Week, October 10/2005, pp. 68-74.
(4) RAWE, Julie, "Why Your Boss May Start Sweating the Small Stuff", Time, March 20/06, pp. 80
(5) ROSENZWEIG, Philip M., "Managing the New Global Workforce: Fostering Diversity, Forging Consistency", submission for European Management Journal, 1998.
(6) GREENWALD, Bruce and Judd Kahn, "All Strategy is Local", Harvard Business Review, September/05, pp. 95-104.
(7) ROSEN, Robert, "Global Literacies", Simon and Schuster, New York, 2000.
(8) McGREGOR, Jena, "Dawn of the Idea Czar", Business Week, April 10/06, pp. 58-59
(9) "Meet the Innovation Gurus", Business Week, August 1/2005, pp. 76-79
(10) SEDEN, Larry and Ian C. MacMillan, "Manage Customer-Centric-Innovation – Systematically", Harvard Business Review, April 2006, pp. 108-116.
(11) BRADY, Diane, "Ideas that Bloom", Business Week SmallBiz Spring 2006, pp. 47-53.

C. Delia Contreras is the production leader for the Dow's IPA / VAM plants in Texas City (TX), phone 409-948-5588; she has 18 years of diversified chemical engineering experience. Before her current role she was production leader for the Dow's CPE plant in Plaquemine (LA) and previous to that she was Work Process Leader for the Dow Acrylics, Solvents & Intermediates businesses and had responsibility for the effective implementation of work processes for 21 plants around the world. She has also worked in Malaysia and Colombia, including leadership positions. She holds a B.S. in Chemical Engineering from Universidad Industrial de Santander (Colombia) and has co-authored other six papers on different subjects. She is the recipient of the 2006 HENAAC's (Hispanic Engineers' National Achievement Award Corporation) Professional

Achievement Award and is a member of the Society of Women Engineers and the Society of Hispanic Professional Engineers.

F. Bravo is a lead process engineer for the Dow Chemical Company, phone 979-238-3253. He has 22 years of experience in the chemical, petrochemical, refining and polymer industries and is co-author of several articles on different subjects. He has had project leadership roles for projects in South America, USA, Italy, Malaysia and China. He holds two degrees in chemical engineering: a B.S. from Universidad Pontificia Bolivariana (Colombia) and a M.S. from Rice University.

167c Business Plans, Investment Planning Models and Financial Risk in Che Capstone Design Classes

Miguel J. Bagajewicz, Chemical, Biological and Materials Engineering, University of Oklahoma, 100 E. Boyd St., T-335, Norman, OK 73019

At the University of Oklahoma, the capstone experience consists of projects conducted in groups of two students, all projects being different.

In this presentation, we will explain how investment planning mathematical programming models as well as financial risk management is incorporated into process design projects.

We will also show how microeconomics and financial risk is used in product design projects.

Examples will be provided to illustrate how close to complex real cases we are able to get.

Demand Modes for Safety Instrumented Functions

Iwan van Beurden
Senior Safety Engineer
exida.com L.L.C.
Sellersville, PA 18960
vanbeurden@exida.com

Rachel van Beurden-Amkreutz
Safety Engineer
exida.com L.L.C.
Sellersville, PA 18960
rachel.amkreutz@exida.com

KEYWORDS

Safety Integrity Level, Safety Instrumented Function, IEC 61508, IEC 61511, PFDavg, PFH, SIL verification, Architectural Constraints, high demand, low demand, continuous mode, demand mode, proof test interval, demand frequency, diagnostic test interval

ABSTRACT

The acceptance of the functional safety standards, IEC 61508 [1] and IEC 61511 [2] / ANSI/ISA 84.00.01:2004 [3] has increased drastically over the last year. Many end-users are requesting manufacturers to provide reliability information on specific products to perform the Safety Integrity Level (SIL) verifications that are required by the functional safety standards.

The SIL performance parameter for Safety Instrumented Functions relates to a range of average Probability of Failure on Demand (PFDavg) values for low demand mode (or demand mode) applications and a range of Probability of a Dangerous Failure per Hour (PFH) values for high demand mode (or continuous mode) applications.

The reliability calculations that are performed as part of the SIL verification allow end-users to optimize their Safety Instrumented Function design, through technology, redundancy, and/or proof test interval selections. Many SIL certified devices have been designed to achieve a low dangerous undetected failure rate which allows for longer intervals between proof tests while still achieving a low PFDavg. The advantages are obvious, less process interruption, less maintenance work, lower lifecycle costs.

The extension of proof test intervals could however result in a SIF application being classified as a high / continuous demand mode application. This paper will explain the rules for classifying an application as low demand or high / continuous demand. It will explain how exida uses three different modes of operation and the motivation for this. Why does an application that seems to be a low demand application, all of a sudden need to be considered as a high demand application? Finally, the paper will also explain the differences between low demand and high demand applications and the consequences for the SIL verification.

INTRODUCTION

The acceptance of the functional safety standards, IEC 61508 [1] and IEC 61511 [2] / ANSI/ISA 84.00.01:2004 [3] has increased drastically over the last year. Many end-users are requesting manufacturers to provide reliability information of specific products to perform the Safety Integrity Level (SIL) verifications that are required by the functional safety standards.

As explained in [4] for compliance with the IEC 61508 or IEC 61511 / ANSI/ISA 84.00.01:2004 functional safety standards, the achieved Safety Integrity Level of a Safety Instrumented Function (for low demand mode applications) is determined by the lower of two calculated Safety Integrity Levels. The achieved SIL is the lower of the SIL based on the average Probability of Failure on Demand and the SIL based on the Architectural Constraints. This is also illustrated in FIGURE 1. The SIL based on the average Probability of Failure on Demand is often represented by SILpfd. The SIL based on the Architectural Constraints is often represented by SILac [5].

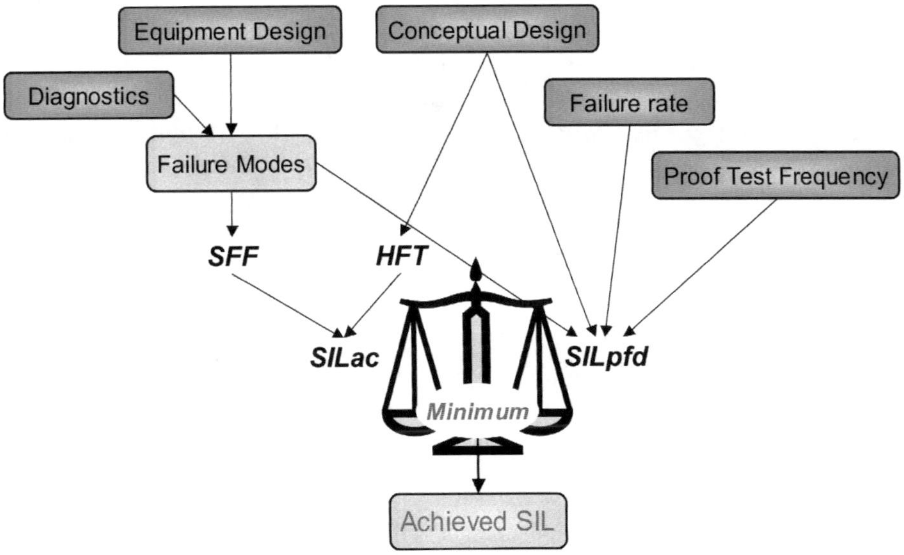

FIGURE 1 SIF ACHIEVED SIL

The determination of the achieved SIL as described in [4] is correct for most process industry cases. It however does not address the different modes of operation that Safety Instrumented Functions can operate in. IEC 61508 distinguishes between the low demand mode and high demand mode of operation. IEC 61511 distinguishes between demand mode and continuous mode Safety Instrumented Functions. The SIL performance parameter for Safety Instrumented Functions relates to a range of average Probability of Failure on Demand (PFDavg) values for low demand mode (or demand mode) applications and a range of Probability of a dangerous Failure per Hour (PFH) values for high demand mode (or continuous mode) applications.

MODES OF OPERATION

As we can see above, the functional safety standards defined various modes of operation for Safety Instrumented Functions. IEC 61508 defines a low demand mode and a high or continuous demand mode, whereas IEC 61511 defines the demand mode and continuous mode applications. Within exida it was concluded that in some cases a third operating mode could be identified. Consequently exida defines three modes of operation. The definitions for the different modes of operation are provided in the sections underneath.

IEC 61508 MODES OF OPERATION

IEC 61508 defines a *mode of operation* as a "way in which a safety-related system is intended to be used, with respect to the frequency of demands made upon it" [1 (3.5.12)]. The mode of operation may be either

- **low demand mode**: where the frequency of demands for operation made on a safety-related system is no greater than one per year and no greater than twice the proof test frequency;
- **high demand or continuous mode**: where the frequency of demands for operation made on a safety-related system is greater than one per year or greater than twice the proof test frequency

In addition, IEC 61508 states that the high demand or continuous mode covers those safety-related systems which implement continuous control to maintain functional safety. It also describes that the target failure measures for safety-related systems operating in low demand mode is the average Probability of Failure on Demand (PFDavg) and for systems operating in high demand or continuous mode is the Probability of a dangerous Failure per Hour (PFH).

There are two main comments that can be made with regard to the IEC 61508 definitions of operating modes,

1. there is no distinction made between high demand and continuous mode, and
2. there is no indication why a frequency of demands of one per year is a "cutoff point"

Clearly if the first mode of operation can be identified using one term, i.e. low demand; it should also be possible to identify the second mode of operation with one term, i.e. high demand. The IEC 61508 committee however thought it necessary to add continuous mode to the second mode of operation description. Why? Arguably there are nuances in the *high demand or continuous mode* operating mode that could result in additional classification. This is an indication that instead of two modes of operation a third mode would need to be defined. This is exactly what exida did as will be described in a subsequent section.

The quantitative boundary between low demand and high demand applications is laid at a frequency of demand no greater than one per year and no greater than twice the proof test frequency. The latter rule, i.e. the demand interval needs to be at least two times greater than the proof test interval, can intuitively be explained. You want to make sure that a proof test is performed at least once before the demand occurs in order to take credit for that proof test. As a demand can occur just before a proof test or just after a proof test, you want to make sure that there are approximately two proof test intervals per demand interval. A similar logic is typically applied when it comes to the determination of the maximum scan time of a safety instrumented function; this is typically half the process safety time.

This leaves the first rule of the quantitative boundary, i.e. why should the demand frequency be no greater than one per year, or in other words why can't a demand interval not be smaller than one year for low demand mode applications? This is a good question. Why can an application not be considered low demand if proof test is done once a month and on average a demand is expected once every 10 months? This scenario may not apply to continuous processes, but can definitely apply to batch processes. We will see that the one year limit does not exist in the exida definitions of demand modes. The omitting of the one year limit is supported by mathematical proof, see [5].

IEC 61511 MODES OF OPERATION

IEC 61511 defines a *mode of operation* as a "way in which a safety instrumented function operates" [2 (3.2.43)]. Two safety instrumented function modes of operation are defined

- **demand mode safety instrumented function**: where a specified action (for example, closing of a valve) is taken in response to process conditions or other demands. In the event of a dangerous failure of the safety instrumented function a potential hazard only occurs in the event of a failure in the process or the BPCS
- **continuous mode safety instrumented function**: where in the event of a dangerous failure of the safety instrumented function a potential hazard will occur without further failure unless action is taken to prevent it

In addition, IEC 61511 states that the Continuous mode covers those safety instrumented functions which implement continuous control to maintain functional safety. IEC 61511 describes that in demand mode applications where the demand rate is more frequent than once per year, the hazard rate will not be higher than the dangerous failure rate of the safety instrumented function and that in such a case, it will normally be appropriate to use the continuous mode criteria. Finally, it also describes that the target failure measures for safety instrumented functions operating in demand mode is the average Probability of Failure on Demand (PFDavg) and for functions operating in continuous mode is the Probability of a dangerous Failure per Hour (PFH).

There are several comments that can be made with regard to the IEC 61511 definitions of operating modes,

1. there is a clear textual definition of continuous mode,
2. there is only one demand mode defined, i.e. no distinction between low and high demand
3. there is no mentioning of the proof test frequency vs. demand frequency ratio, and

4. the demand frequency "cutoff point" of once per year shows up again

The IEC 61511 committee provided us with a clear textual definition of continuous mode applications. The indication we got from IEC 61508 that a third mode needed to be defined instead of a combination of high and continuous was also noticed by the IEC 61511 committee. There is however in this definition no indication as to how continuous mode safety instrumented functions need to be treated differently than non-continuous mode safety instrumented functions. The exida definitions will provide a clear quantitative indication as to when a Safety Instrumented Function should be considered operating in continuous demand and when not.

While we got a clear definition of our "third" operating mode, the IEC 61511 committee decided to leave out the IEC 61508 definitions on low and high demand mode. As described in the previous section there is an intuitively understanding as to why a distinction needs to be made with regard to the ratio of proof test frequency and demand frequency when that ratio gets too close to two (2). As a matter of fact, in its definitions IEC 61511 does not discuss the ratio between proof test frequency and demand frequency at all. It can be mathematically proven that the demand mode needs to be divided into a low and high demand mode [5]. The exida modes of operation define both.

The quantitative boundary for the demand frequency, i.e. a demand cannot be more frequent than once per year for demand mode applications, is also part of the definitions in IEC 61511. The same comment applies here as was made for the IEC 61508 modes of operation.

EXIDA MODES OF OPERATION

Based on the extensive experience of the exida partners and employees within the various industries and functional safety committees, we concluded that a total of three modes of operation should be defined for safety instrumented functions. Similar to IEC 61508 and IEC 61511, exida defines a *mode of operation* as a "way in which a safety instrumented function operates". The safety instrumented function modes of operation defined are

- **low demand mode**: where the frequency of demands for operation made on a safety instrumented function is no greater than twice the proof test frequency;
- **high demand mode**: where the frequency of demands for operation made on a safety-related system is greater than twice the proof test frequency and where the worst case diagnostic test interval is at least an order of magnitude smaller than the demand interval
- **continuous demand mode**: where the frequency of demands for operation made on a safety-related system is greater than twice the proof test frequency and where the worst case diagnostic test interval is not an order of magnitude smaller than the demand interval

The definitions are very clearly oriented towards the behavior of the safety instrumented function with regard to the ability to perform a proof test and the ability to execute automatic diagnostics within the demand interval. Based on the ratio of demand frequency to proof test frequency and of demand frequency to diagnostic time frequency, the table shown in FIGURE 2 can be derived. This table defines what probability of failure measure, PFDavg or PFH applies for each of the three operating modes. It also indicates if credit can be taken for proof testing and if credit can be taken for automatic diagnostics. Using the guidelines in this table the SIL verification calculations can easily be performed.

exida Modes of Operation		
Low Demand	**High Demand**	**Continuous Demand**
Use PFDavg table	Use PFH table	Use PFH table
Take credit for proof testing	No credit for proof testing	No credit for proof testing
Take credit for automatic diagnostics	Take credit for automatic diagnostics	No credit for automatic diagnostics

FIGURE 2 EXIDA MODES OF OPERATION

DETERMINE MODE OF OPERATION

The reliability calculations that are performed as part of the SIL verification allow end-users to optimize their Safety Instrumented Function design through technology, redundancy, and/or proof test interval selections. Many SIL certified devices have been designed to achieve a low number of dangerous undetected failures which allows for longer intervals between proof tests while still achieving a low PFDavg (this assumes a low demand mode Safety Instrumented Function). The relation between dangerous undetected failure rate and proof test interval is shown in the simplified equation for a single (1oo1) system.

$$PFDavg = \frac{\lambda^{DU} \cdot TI}{2} + \lambda^{DD} \cdot MTTR \qquad \textbf{EQUATION 1}$$

The advantages of a larger proof test interval are obvious, less process interruption, less maintenance work, lower lifecycle costs, etc. The lengthening of the proof test interval could however result in the SIF application being classified as a high demand or continuous demand mode application. Based on the exida description of safety instrumented function demand modes the following algorithm can be derived to determine in what demand mode a function is operating.

```
If  DemandInterval  >  ( 2 * ProofTestInterval )  then
        Low Demand
else
        if  DemandInterval  >  ( 10 * DiagnosticTestInterval )  then
                High Demand
```

else

Continuous Demand

end if

end if

The Diagnostic Test Interval used in the algorithm refers to the worst case diagnostic test interval of all equipment items used in a Safety Instrumented Function. Manufacturers of equipment items used in Safety Instrumented System applications are required to publish this number in the product's safety manual. So in order to determine the appropriate diagnostic test interval to consider in the mode of operation algorithm one needs to compare all worst case diagnostic test intervals of the equipment items in the Safety Instrumented Function and select the longest interval.

WHAT MODE OF OPERATION APPLIES TO MY SIF?

If we keep the table shown in FIGURE 2 in our minds, we can easily see what the resulting mode of operation means for our SIL verification given a set of assumed failure rates, diagnostic time intervals, demand rates, and proof test intervals. Assume that we are shooting for a target SIL of SIL 2.

For the sake of clarity assume that a single set of failure rates represents an entire Safety Instrumented Function [6].

λ^{SD} = 1550 FIT
λ^{SU} = 325 FIT
λ^{DD} = 1750 FIT
λ^{DU} = 130 FIT

Mean Time To Restoration (MTTR) = 8 hrs

Worst case Diagnostic Test Interval = 24 hrs

If a demand on the safety instrumented function occurs once every 6 years and we perform a proof test every year, we are obviously operating in a low demand mode. Consequently PFDavg needs to be considered and we take credit for proof test and automatic diagnostics. Using the formula from EQUATION 1, a PFDavg of 5.8E-4 results, see underneath.

$$PFDavg = \frac{130 \cdot 10^{-9} \cdot 1 \cdot 8760}{2} + 1750 \cdot 10^{-9} \cdot 8 = 5.8 \cdot 10^{-4}$$ **EQUATION 2**

Given that for a SIL 2 application the PFDavg value of the entire safety instrumented function needs to be lower than 1.0E-2 we are easily making the SIL 2 target[1].

So why not extend our proof testing to lets say 10 years? If we rerun the calculation with a 10 year proof test interval, a PFDavg of 5.7E-3 results as shown underneath.

---

[1] In this example SIL verification we are purposefully excluding the architectural constraints considerations and assume we are not limited by the SILac irrespective if we are dealing with Type A or Type B devices.

$$PFDavg = \frac{130 \cdot 10^{-9} \cdot 10 \cdot 8760}{2} + 1750 \cdot 10^{-9} \cdot 8 = 5.7 \cdot 10^{-3}$$ **EQUATION 3**

This PFDavg is still below the SIL 2 limit, so one would think that we are all done. However, looking back at the mode of operation algorithm the safety instrumented function is not operating in the low demand mode anymore, after all the demand interval of 6 years is smaller than the proof test interval of 10 years. Consequently PFDavg does not apply anymore!

Re-evaluating the mode of operation using the proof test interval of 10 years and the demand interval of 6 years, we are running in either the high or continuous mode. Since the worst case diagnostic test interval is only 24 hrs the safety instrumented function is operating in the high demand mode. A very straight forward simplified equation can be derived to calculate the Probability of a dangerous Failure per Hour (PFH) for a single (1oo1) system.

$$PFH = \lambda^{DU}$$ **EQUATION 4**

Using this equation we can calculate what the Probability of a dangerous Failure per Hour is for our safety instrumented function, PFH = 1.3E-7 / hr, see underneath.

$$PFH = 130 \cdot 10^{-9} = 1.3 \cdot 10^{-7} / hr$$ **EQUATION 5**

For a SIL 2 application the PFH value of the entire safety instrumented function needs to be lower than 1.0E-6 / hr. So given our PFH calculation we are still making the SIL 2 target[2]. Note that additional requirements apply for the safety instrumented function if that function is operating in high demand mode rather than low demand mode. These will be described in a subsequent section.

The assumptions made so far can apply to most process industry applications. However when we consider a machinery application a demand on a safety instrumented function may occur very frequently. A demand could occur every hour, for example if an operator inadvertently intrudes a specific area. A demand could occur even more frequent in situations where the safety instrumented function implements continuous control. In these cases the demand interval will not be an order of magnitude larger than the worst case diagnostic test interval, it is in fact smaller, and the safety instrumented function will be operating in continuous demand per the mode of operation algorithm.

For a continuous demand mode application we can see in FIGURE 2 that the Probability of a dangerous Failure per Hour applies but we cannot take any credit for automatic diagnostics. Therefore we will need to consider all dangerous failures undetected. Updating the PFH formula leads to

$$PFH = \lambda^{DD} + \lambda^{DU}$$ **EQUATION 6**

[2] The architectural constraints concept / SFF concept only applies to low demand mode applications.

Using this equation we can calculate what the Probability of a dangerous Failure per Hour is for our safety instrumented function, PFH = 1.9E-6 / hr, see underneath.

$$PFH = 1750 \cdot 10^{-9} + 130 \cdot 10^{-9} = 1.9 \cdot 10^{-6} / hr \qquad \textbf{EQUATION 7}$$

This calculated PFH is not lower than the SIL 2 limit of 1.0E-6 / hr, so in this case we do not meet our SIL 2 target, we only meet SIL 1.

WHAT IS THE SIF DEMAND FREQUENCY?

A common misconception is that the demand frequency on the safety instrumented function can easily be derived from the SIL Selection. After all the demand frequency is input to the Hazard Matrix and Risk Graph. In addition the Layer of Protection Analysis (LOPA) yields a frequency as well. It should be noted however that the demand frequency used in the Hazard Matrix and Risk Graph analysis is representing the frequency at which the hazard occurs that the safety instrumented function is protecting against. This usually takes various safeguards like pressure relief devices into consideration. This is more easily seen from the LOPA diagram in FIGURE 3.

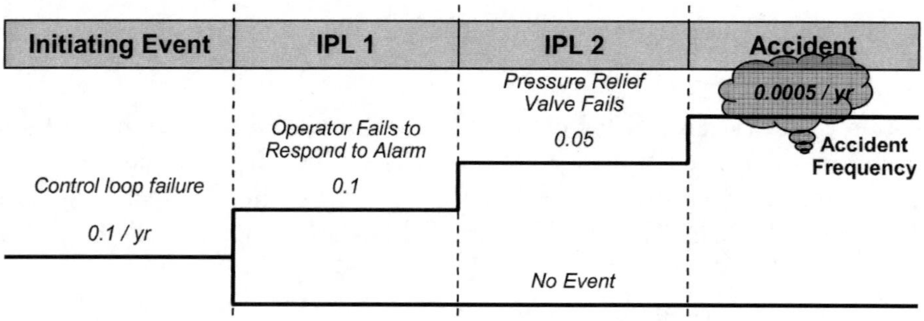

FIGURE 3 EXAMPLE LOPA DIAGRAM

As one can see the hazard will occur if the initiating event occurs, the operator fails to respond to an alarm, and the pressure relief valve fails. The accident frequency can simply be calculated by multiplying the initiating event frequency with the two independent layers of protection probabilities of failure on demand.

The initiating event will however lead to a demand on the safety instrumented function if the initiating event occurs and if the operator fails to respond to the alarm. The safety instrumented function will most likely be executed before the pressure relief valve performs its function. The adjusted LOPA diagram in FIGURE 4 shows how the frequency of demands on the safety instrumented function can be determined.

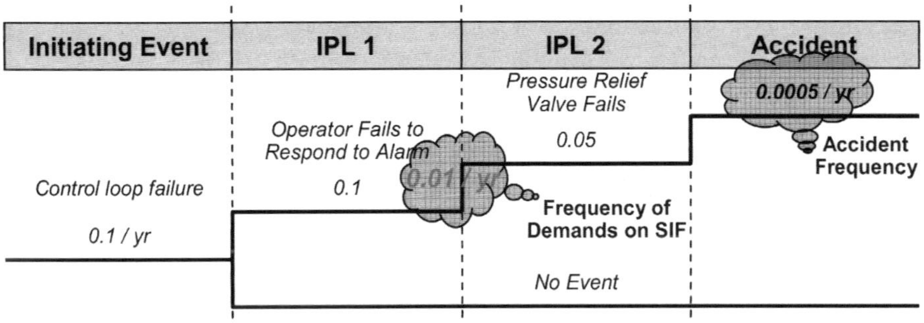

FIGURE 4 EXAMPLE LOPA DIAGRAM SHOWING SIF DEMAND FREQUENCY

APPLICATION DIFFERENCES BECAUSE OF MODE OF OPERATION

Both IEC 61508 and IEC 61511 have specific requirements regarding the behavior of safety instrumented functions based on their mode of operation. The distinction is basically between the low demand mode of operation on one hand and the high and continuous demand mode of operation on the other hand. The differences in how safety instrumented functions need to be applied based on their mode of operation will be discussed in this section.

The application difference because of mode of operation relates to the behavior of the safety instrumented function upon detection of a fault. The detection can be by (automatic) diagnostic tests, proof tests or by any other means. Two situations are considered; the subsystem of the safety instrumented system that the fault occurs in can or cannot tolerate a single hardware fault. In the first situation where the subsystem can tolerate a hardware fault a Hardware Fault Tolerance (HFT) greater than zero exists, in the latter situation the Hardware Fault Tolerance of the subsystem is zero.

SIF WITH HFT GREATER THAN ZERO

If a dangerous fault is detected in a subsystem with a Hardware Fault Tolerance greater than zero, i.e. the subsystem can tolerate a single hardware fault, there are two options for the behavior of that subsystem:

a) a specified action to achieve or maintain a safe state shall result; or
b) the subsystem continues safe operation of the process while the faulty part is repaired. If the repair of the faulty part is not completed within the mean time to restoration (MTTR) assumed in the calculation of the probability of random hardware failure, then a specified action shall take place to achieve or maintain a safe state.

The safe state should be specified in the safety requirements specification.

For the situation that the safety instrumented function has a hardware fault tolerance greater than zero no distinction is made between the behaviors of the safety instrumented function based on the mode of operation.

SIF WITH HFT EQUAL TO ZERO

If a dangerous fault is detected in a subsystem with no Hardware Fault Tolerance, i.e. the subsystem cannot tolerate a single hardware fault, a distinction is made between the modes of operation of the safety instrumented function. If the subsystem is operating in the low demand mode of operation, there are two options for the behavior of that subsystem:

a) a specified action to achieve or maintain a safe state shall result; or
b) the faulty subsystem needs to be repaired within the mean-time-to-restoration (MTTR) period assumed in the calculation of the probability of random hardware failure. During this time the continuing safety of the process shall be ensured by additional measures and constraints. The risk reduction provided by these measures and constraints shall be at least equal to the risk reduction provided by the safety instrumented function in the absence of any faults. If the repair is not undertaken within the specified mean time to restoration (MTTR) then a specified action shall be performed to achieve or maintain a safe state.

The additional measures and constraints shall be specified in the SIS operation and maintenance procedures. Again the safe state should be specified in the safety requirements specification.

The requirements for a safety instrumented function that is operating in a low demand mode of operation are almost identical to the requirements for a safety instrumented function with a hardware fault tolerance greater than zero.

If the subsystem with no Hardware Fault Tolerance that the fault is detected in is operating in the high or continuous demand mode of operation, there is only one option for the behavior of that subsystem:

a) a specified action to achieve or maintain a safe state shall result

The specified action (fault reaction) required to achieve or maintain a safe state shall be specified in the safety requirements specification. The total time to detect the fault and to perform the action shall be less than the time for the hazardous event to occur.

This is the main difference between the applications of safety instrumented functions in different modes of operation. Safety instrumented functions that run in high or continuous demand mode cannot continue to operate once a fault is detected and there is no hardware fault tolerance.

In addition to the specified behavior upon fault detection, there is also a requirement regarding the diagnostic test interval for safety instrumented functions operating in either the low or high or continuous demand mode of operation.

For safety instrumented functions operating in the high or continuous demand mode the diagnostic test interval of any subsystem having a hardware fault tolerance of zero shall be such that the sum of the diagnostic test interval and the time to perform the specified action (fault reaction) to achieve or maintain a safe state is less than the process safety time.

Besides, and this is valid for both low and high or continuous demand applications, if the diagnostic test interval is not at least and order of magnitude smaller than the demand interval the diagnostic test interval needs to be directly included in the reliability modeling.

CONCLUSION

The performance based concept of the functional safety standards allow end-users to optimize their Safety Instrumented Function design, from a technology, redundancy, and proof test interval perspective given the reliability calculations that are performed as part of the SIL verifications. Since many SIL certified devices have been designed to achieve a low dangerous undetected failure rate users are able to specify longer intervals between proof tests while still achieving a low PFDavg. The objective being less process interruption, less maintenance work, lower lifecycle costs, etc.

When evaluation safety instrumented functions three different modes of operation can be identified;

- **low demand mode**: where the frequency of demands for operation made on a safety instrumented function is no greater than twice the proof test frequency;
- **high demand mode**: where the frequency of demands for operation made on a safety-related system is greater than twice the proof test frequency and where the worst case diagnostic test interval is at least an order of magnitude smaller than the demand interval
- **continuous demand mode**: where the frequency of demands for operation made on a safety-related system is greater than twice the proof test frequency and where the worst case diagnostic test interval is not an order of magnitude smaller than the demand interval

For each of these modes of operation there are guidelines with regard to the SIL verification as to if credit can be taken for proof testing and if credit can be taken for automatic diagnostics. Additional requirements with regard to the application of the safety instrumented functions exist. For any function that is considered to operate in the high or continuous demand mode and that has no hardware fault tolerance, continued (degraded) operation once a fault is detected is not permitted. The safety instrumented functions will need to achieve or maintain a safe state. Applications that operate in low demand mode can continue to operate in the degraded state as long as the fault is repaired within the mean-time-to-restoration that was considered in the reliability calculations for the specific safety instrumented function.

REFERENCES

[1] IEC 61508, Functional safety of electrical / electronic / programmable electronic safety-related systems, 2000, International Electrotechnical Commission, Geneva, Switzerland

[2] IEC 61511, Functional safety: Safety Instrumented Systems for the process industry sector, 2003, International Electrotechnical Commission, Geneva, Switzerland

[3] ANSI/ISA 84.00.01:2004 (IEC 61511 Mod), Functional Safety: Safety Instrumented Systems for the Process Industry Sector, 2004, Instrument Society of America, Research Triangle Park, NC, USA

[4] van Beurden I., What does Proven In Use imply?, Technical Papers Collection ISA volume 454, ISA 2004, Houston, Texas, USA

[5] exida Safety Integrity Level verification tool, SILver, part of exSILentia, exida.com LLC, Sellersville, PA, USA

[5] Bukowski J., Incorporating Process Demand into Models for Assessment of Safety System Performance, Annual Reliability and Maintainability Symposium 2006 Proceedings, Newport Beach, CA, USA.

[6] Safety Equipment Reliability Handbook, 2nd edition, 2005, exida.com L.L.C., Sellersville, PA, USA, ISBN-13 978-0-9727234-1-1

ABBREVIATIONS AND DEFINITIONS

HFT	Hardware Fault Tolerance
IEC	International Electrotechnical Commission
LOPA	Layer of Protection Analysis
MTTR	Mean Time To Restoration
PFDavg	Average Probability of Failure on Demand
PFH	Probability of a dangerous Failure per Hour
SIF	Safety Instrumented Function
SIL	Safety Integrity Level
SILac	Achieved Safety Integrity Level based on Architectural Constraints
SILpfd	Achieved Safety Integrity Level based on the Safety Instrumented Function's PFDavg
SILpfh	Achieved Safety Integrity Level based on the Safety Instrumented Function's PFH
SIS	Safety Instrumented System
λ^{SD}	Safe Detected failure rate
λ^{SU}	Safe Undetected failure rate
λ^{DD}	Dangerous Detected failure rate
λ^{DU}	Dangerous Undetected failure rate
MTTR	Mean Time To Restoration
TI	Test Interval

Abstract #102115

Dynamics of Global Market & Effective Communication Based Leadership Style

Syamal K. Poddar, Fuels and Petrochemicals Division/Group 16, Poddar and Associates, P.O. Box 890461, Houston, TX 77259-0461

Globalization and its impact:

Globalization in business is not a new phenomenon. It has been in existence for centuries. In the ancient world there were businesses beyond each country's boundary. However, they were primarily trading businesses and they were mostly between parts of the world in the same region. The changes were happening in slow speed and communication linkages were rather poor. Because the process was slow, mutual understanding and integration was taking place incrementally in slow steps. In a thermodynamic sense it was following a reversible path. With industrialization, the process of globalization has hastened, particularly after World War II, and it has been accelerating ever since. Most recently, the process of globalization has been moving even faster. Thermodynamically such changes have become irreversible. Emerging players like China and India, with approximately one third of world's population. are appearing in the forefront of this ever expanding global homogenization process and aspiring to be major players. The trade barriers between various countries are being lifted. The market competition is expanding. Over the next 20 to 30 years the global market of about $10 trillion could transform to over the $100 to 150 trillion range. The "safe harbors" from foreign market competitions are disappearing.

This globalization phenomenon is impacting across all types of businesses including goods and services types. With technology innovations there is a convergence of diverse business sectors e.g., computers, information technology, medical devices, pharmaceuticals, biotechnology, nanotechnology, energy sources & their utilization and power generation & transmission to name a few.

Globalization is creating a multitude of opportunities. Some of the notable ones are expansion of trade, increase of foreign direct investment, technology convergence as well as information technology innovation. With these opportunities, the global business leaders are faced with a lot of rather complex challenges in various sectors of business which are organizational as well as technological and operational in nature. A few fundamental challenges are: cultural differences and diversity, less competent human capital and substandard economic infrastructure.

To face these challenges and to harness the multitude of opportunities, competent and focused yet flexible leadership with specific traits is a must.

Leadership:

There is a fundamental difference between management and leadership. "Managers are people who do things right and leaders are people who do the right thing" Vision is a unique trait of any true leader by which he overcomes the barriers to change. Effective leadership will be able to move the organization from the current state to a future state satisfying the impact of these step changes. A leader is able to instill a common vision in employees across the boundary by absorbing available cultural diversity as a strength and not setting aside such diversity as a weakness. A true leader instills confidence in the organization by empowering the employees. An organization driven by an effective leader translates his vision to reality. This translation is manifested by an effective 360 degree communication between leaders and their followers.

The two important leadership traits are credibility and commitment fueled by trust and. integrity. The complexity of the business world has caused leaders to think and act differently. They are there to make a difference and to create changes to cope with the challenges. The global leadership is no longer the traditional hierarchical management style driven by linear thinking and guided by implementation of incremental strategy. The changes are in various fronts of national and global economies.

Getting the message across uniformly at all levels is the key to success. This is achieved through establishing a common goal (vision) and creating alignment among all employees. Trust is the glue that keeps the organization moving forward and effective leadership is the catalyst which helps build that trust. Trust creates accountability, dependability, integrity, predictability and the identity of any organization. The effective leader pulls people together by attracting and energizing and motivating them towards a common goal. Effective leaders are super listeners. Above all, a true leader has common sense.

Communication

Most often communication is taken for granted. Effective communication is the hardest to master. An average worker spends more than 80% of time communicating. Poor communication results in more than 25% of work-related mistakes. Written communication is much harder than verbal communication. Non verbal communication plays a very important role. Communication can be within a group or between two individuals

Communication is a two way process. Responding and/or listening are equally important. Characteristics of an effective communication is to transmit a message. There are two sides of a Communication Linkage: 1) Transmitter: Originator of a communication (Speaker, Writer) and 2) Receiver: Target individual or a group for a given communication (Listener, Reader) For effective communication, both sides of the linkage should be at the optimum level of tuning. Communication linkage breaks if: a) either transmitter or receiver overpowers the other and/or; b) either transmitter or receiver significantly underperforms. We spend years learning how to read and write and, to some extent how to speak, but do we get any formal training how to listen to others, to really and deeply understand them? The answer is emphatically "NO". Therefore one of the most important communication skills is totally ignored. Listening skill is the most neglected part of communication training. Yet, listening and, more importantly, active listening, is a critically important item in a communication linkage.

The main driver for any communication is information sharing and information is activated by any form of uncertainties. Uncertainties induce leaders to think and this thinking process creates innovations and change. In thermodynamic terms, uncertainties and chaotic situations, be it in a physical world or in an organization, cause increase in Entropy. To bring alignment, systematization, and structure to an organization, minimization of this Entropy is required. This process requires input of external energy, which is basically the leaders' visionary set of actions. This is achieved by effective communication of necessary information coupled with necessary decision and action. Therefore it is important to recognize the importance of the leaders' effective communication skill which drives a non equilibrium state caused by external changes to an equilibrium state, which is an organization inline with the vision.

Non-verbal communication:

Non-verbal communication is communication without uttering a word. Verbal communication is organized by language; non verbal communication is not. It is very prevalent during a face-to-face or in a group setting communication where 90% of communication is accomplished non-verbally. Non-verbal signals, often called body language, are easy to miss. Often the verbal and nonverbal messages contradict causing confusion. Body language, carefully observed and interpreted, can tell a lot about what others are thinking or feeling. Nonverbal communication is learned and practiced often on a subconscious level. Body language is open to misinterpretation just as verbal communication is. Interpretation of body language being subject to one's lifestyle, family, cultural background, and other factors, understanding of non-verbal communication and body language is extremely important in effective global communication. Non-verbal communication includes facial expressions, eye contact, tone of voice, body posture and motions, and positioning within groups. In case of any mixed (confused) message it is better to depend on non-verbal rather than verbal communication because verbal words can often be manipulated. On the other hand, body language is hard to control. Therefore, body language never lies. Paying attention to non-verbal clues can bring a lot of clarity during face-to-face communication.

Cross-Cultural Communication:

Global leaders need to be culturally sensitive and globally minded. Creating effective human capital from a diverse cultural background is becoming more of a challenge than monetary capital. Global leaders are called to managing accelerating changes and differences and they need to willingly re-examine and alter personal attitudes and perceptions. Proficiency in cross-cultural communication is of tantamount importance. The leaders are to create cultural synergy whereby the leaders are to emphasize similarities and common concerns and integrate differences to enrich organizational strength. As progress is made in advancing a multi-culture based leadership style, a global culture with some common characteristics is being created. The impact of mass media and telecommunication, the internet and emails are bringing homogenization between cultures in various parts of the world. Transnational business strategies are emerging to cope with globalization. For the leaders in global organizations sensitivity and understanding of the transnational cultural differences across the board, including their customers, competitors, suppliers, business or alliance partners and government officials, are basic requirements. This is only achievable if the global leaders become more proficient in cross cultural communication. The global organization's objective is to create an integrated approach to establish a unified goal. It imbibes a strong sense of team work across the world. Culturally sensitive and skilled leaders who value diversity as strength and not weakness, are the backbone of any successful sustainable global organization. Cross cultural communication is the most important basic skill for any successful global leaders There is no unified approach to master the cross cultural communication skill other than understanding each and individual culture of the countries such a global organization is targeting as its domain of operation. The key element of success in a multi-cultural communication setup is proficiency in non-verbal communication.

There is going to be a significant change in the mind set of business leaders in terms of how they think about their industry sectors and their management tasks. The structures of companies are changing from domestic to multi-domestic or multinational as well as from local to regional to multi-regional to global. The organizations from various countries are becoming more interdependent requiring their business leaders to communicate more effectively.

Concluding Remarks:

The world market is expanding and barriers between countries are dissolving. As a result, the demand of the world market is no longer influenced by market players of developed countries alone. The market players from developing and under-developed countries are becoming more influential. Doing business in one's domestic front alone is no longer a sound sustainable business strategy.

Two distinct yet complementary forces are shaping the world market: technology - innovation and globalization. The convergence of these two forces is challenging global companies to: 1) offer higher quality, more reliable and more standardized products and services at a lower cost to thereby improve profitability; and 2) establish more market share. Innovations and product development extend world-wide and are expanding rapidly. Technological advances and competitive market pressures are making new product life-cycles shorter than ever before. The rate of assimilation of these advancements is decreasing, and product development costs are skyrocketing. Thus business dynamics are getting more complex and challenging than ever before. As this metamorphosis is taking shape and continuing, it is forcing the past equilibrium of these dynamic forces towards a new one.

With such transformations taking place, in order to manage these dynamics the leadership style itself is undergoing substantial changes. Such changes include a better understanding of cultural differences, business practices, business ethics and styles of the multicultural business players. Pro-active planning is becoming of paramount importance in understanding such transformations. Such understanding helps to identify necessary action steps and their subsequent implementation. It is becoming necessary to fine-tune the organizational structure to effectively achieve global presence with a local touch.

Among all the important traits of a successful leader, the most invaluable one is his/her skill at effective communication. Coupled with the necessary technical and/or business knowledge and understanding, effective communication skills are essential for business leaders to face this ever-changing global marketplace and to establish sustainable business success. Figure 1 below schematically describes a successful leadership style

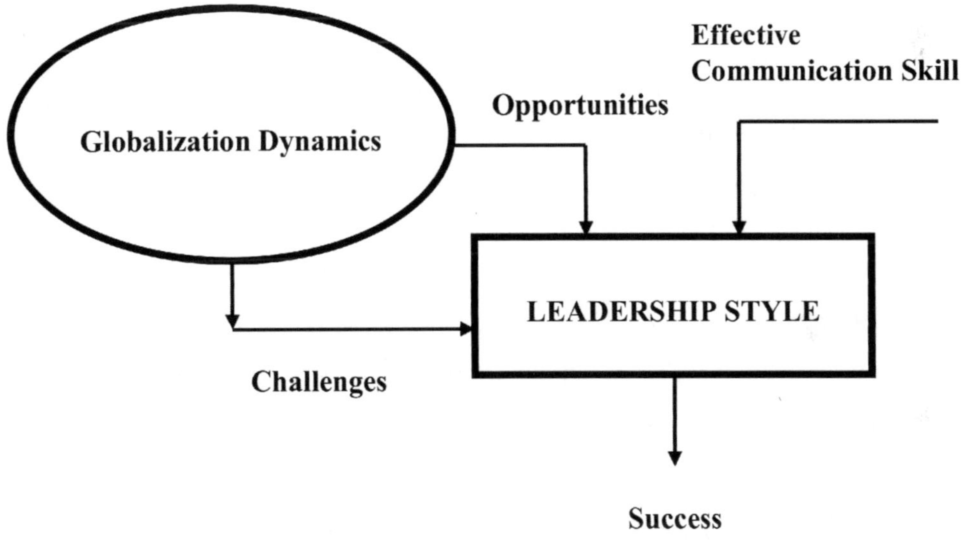

Figure 1
Effective Communication Based Leadership Style Driving Success in Global Market

"Communication is the most important skill in life" Stephen Covey
"The total effectiveness of leaders rises or falls in direct proportion to their face-to-face communication skills" Joe Batten

Managing Conflict

Dr. Eldon R. Larsen, PMP, Marshall University, South Charleston West Virginia

Ryland W. Musick, Jr., P.E., West Virginia Division of Highways

Conflict management is not a topic of rocket science, with all sorts of facts and figures. It is a topic based on morality, communication, care and concern, and good common sense. It is how managers and employees respond during conflicts that determines the success of conflict management. Focusing on the principles provided in this paper may give assistance to all involved in conflict. Remember, every person and situation is different.

This paper is about managing conflict, whether personal or interpersonal. We have chosen the wording carefully here. We speak of *managing* conflict, rather than *resolving* conflict, because (1) you will likely not be able to resolve all conflicts; (2) you don't necessarily need or want to resolve all conflicts—in doing so, for example, you might become much too oppressive and authoritarian and stifle creativity and enthusiasm; however, (3) you must *manage* conflict—that is part of your job.

We will discuss what conflict is, some good and bad effects of conflict, the basic reasons for conflict, and some suggestions for managing conflict.

Conflict Basics

In order to approach conflict management, let's understand some basics.

First, we will define *conflict* to be differing points of view, or actions, which *seem* at least in part to be contradictory, in disagreement, or out of harmony. We use the word *seem*, because our experience shows that very often people think they disagree, when in reality they don't—they just haven't communicated well enough to realize that they are saying the same thing in different words and from differing perspectives. It is not uncommon for people feeling conflict to realize they actually agree after a fair amount of heart-to-heart discussion. This happens a lot in marriage. On the other hand, people often do truly disagree. In any case, whether we truly disagree or not, it is our *perception* of disagreement that is important, because for people their perceptions to them are reality until that perception gets changed or resolved.

Second, conflict is not the same as contention. Although conflict might lead to contention, and both conflict and contention can occur at the same time, *conflict* has to do with *perceived differences*, while *contention* has to do with *emotions* and possibly aggression. I can differ with you without being angry with you; or alternatively I could be angry with you—the bottom line here is that you and I have a choice whether to get emotional or not! This is fundamental to conflict management: people have the freedom to choose their responses to stimuli. Additionally, proactive people *take responsibility for their choices and actions*. For an excellent discussion of these principles, we refer the reader to Stephen Covey's first habit, *Be Proactive*, in his book *The Seven Habits of*

Highly Effective People. In general, conflict can be either constructive or destructive, but contention is almost always destructive and needs to be avoided whenever a better choice can be found—and usually a better choice can be found.

Third, conflict is part of life. We are different from each other. We have different backgrounds, different training, different experience, different values, different ways of thinking, different understandings of words, different physical skills and limitations, different health challenges, and so on and so on. These differences are not bad; in fact they can be very, very good. Without differences there would be very little opportunity for growth and increased understanding. Without differences in people, every committee of consisting of x number of people could be replaced with one person and x-minus-one number of mirrors in which all the mirrored images just vote the same way as the one person facing the mirrors—this is no good. There is a reason for the old saying that two heads are better than one—because it is true.

Fourth, there is no best way or style to manage conflict—it depends on the situations, needs, and people involved; so every case is different. On the other hand, there are principles and concepts that can allow us to improve dramatically how we manage conflict. The principles don't really change; instead the application of the principles changes according to the situations, needs, and people involved.

Some good and bad aspects of conflict

Although it sometimes might not seem like it, there are often some good aspects of conflict. For example, conflict has the potential to

→ Increase communication

→ Promote creativity

→ Improve procedures

→ Increase understanding

→ Help in the path toward synergy

→ Be a motive for change

→ Enhance personal growth

→ Improve relationships

→ Improve productivity

→ Enhance ownership

On the other hand, conflict also has the potential to

→ Decrease communication

→ Decrease creativity

→ Create procedural difficulties

→ Decrease understanding

→ Hinder synergy

→ Block change

→ Hinder personal growth

→ Worsen relationships

→ Worsen productivity

→ Decrease ownership

The differences in the two lists above are determined to a large degree how the people involved in the conflict, including peers, supervisors, and others, actually handle the conflict.

Basic Reasons for Conflict

In order to understand better how to manage conflict, we first need to understand the main causes or reasons for conflict. A surprisingly large majority of conflicts can be categorized as follows. Listed in order of increasingly-difficult-to-resolve-or-to-manage, most conflicts are due to one or more of the following.

→ Perceived differences in ***facts*** (easiest to resolve)

→ Poor or insufficient ***communication***

→ Perceived differences in ***methods***

→ Perceived differences in ***objectives***, or

→ Perceived differences in ***values*** (most difficult to resolve)

Facts. Conflict due to perceived differences in facts is generally relatively easy to manage, because facts usually can be checked and verified. We can talk to experts, do measurements, do a literature searches, or take other steps to find what is fact and what is not, or what is just an opinion or a theory. Note that even at this relatively easy level, some conflicts might not be able to be *resolved*; but at least we, or the people involved, can learn enough information to help us rather easily *manage* this kind of conflict.

Communication. Conflict due to poor or insufficient communication is very common, and it could be the most common cause of conflict. Communication is a two-way or multi-way process—*communication* is a French word, whose roots (*comme*, *un*, and *ication*) combine to mean "the process of becoming like one or as one."

Because communication is a process, it takes time, patience, and effort. Conflict due to miscommunication is often caused by either misunderstanding or the lack of patience to follow good communication principles. These can cause the conflict to be focused on the wrong person, thing, or idea. If conflict is misattributed or misplaced, it can become both irresolvable and unmanageable. So, good communication is vital.

Although good communication often is difficult, it is actually one of the easiest ways to resolve or to manage conflict; but this is true only if one follows the concepts and principles fundamental to good communication. The bottom line here is that before we do anything else to manage conflict, we need to be applying the principles and concepts involved in good communication. By doing this, conflict often vanishes. Even if the conflict doesn't vanish, its source will be much clearer and better understood.

Methods. Conflicts involving methods generally have to do with perceived differences in how or what procedures ought to be followed to achieve given objectives. Management of *methods conflict* is more difficult that just searching out the truth about facts, but still we can manage this type of conflict in general through a logical analysis and evaluation of various alternative approaches. Depending on the complexity of the situation, such an analysis might be rather simple, or at the other end of the complexity spectrum it could involve following a structured approach using well-established decision-analysis techniques.

Objectives. Conflicts having to do with objectives are more difficult to resolve than those based on facts, communication, or methods. People have organizational objectives, family objectives, personal objectives, and sometimes objectives associated with multiple organizations. Thus, there is much opportunity for conflict to arise.

Many of the more successful organizations have learned that they must pay attention to people's personal and family objectives in addition to organizational objectives, and they must try to harmonize these as much as is reasonably possible.

There can even be intra-organizational objectives conflicts—for example, the research and development department might have some objectives around discovery of new ideas and products that might have high technical risks, whereas the manufacturing department just wants products that are guaranteed to work with available equipment. They both have objectives important to them, and conflict is very possible.

When working to manage *objectives conflict*, the people involved can benefit by trying to take a systems approach. This means that people need to try to look at how the various alternatives affect the entire organization, not just one or a few pieces of the organization. Sub-optimization occurs when we make decisions based only on what is best or optimum for one person or one part of the organization, rather than what is best for the whole.

In addition, whenever there are conflicting objectives among people, parts of organizations, or multiple organizations, Stephen Covey's concept of *looking for Third Alternatives* is essential. As Covey states, *it's not my way or your way, it's a better way, a third alternative.* By looking for new objectives that complement and enhance the objectives we originally had in conflict, we truly can get synergistic solutions—solutions or objectives that are better than either of the original objectives or their sum.

Values. Value-based conflict is the most difficult to manage because values go close to the hearts of individuals and their beliefs. Values tend to be sacred, or nearly so, to people and we are generally unwilling to allow anyone to get even close to them. One thing is certain to the authors: you and I cannot change another person's values—only that person can change his values, and it must be a voluntary choice. Because value-based conflicts are so very difficult to manage, the best approach is usually to try to re-construct the conflict into one that is either objective-based or method-based (see above) and then to manage the conflict at that level.

Some Considerations When Managing Conflict

There is not room in this paper to exhaustively cover all the many approaches and styles for managing conflict, nor is that the purpose. What we want to do is to gain a basic understanding of the principles and approaches involved, so that we can use these to help us deal with the many possible situations that can and will occur.

Who Owns the Problem? One of the first steps to take when conflict arises is to identify who *owns* the problem. This means that we need to identify who is primarily taking responsibility for the decisions that are needed for the source of the conflict.

For example, if a project manager is working with a team member who is not ready to take on the responsibility of making decisions on his own, and that team member has to be told each task to do, then the project manager will likely own the resulting project problems that arise involving that employee. In this case, it would likely be the project manager's responsibility to make the decisions needed to manage the conflict. This does not mean that the project manager does not listen to and communicate clearly with the team member; it simply means that the ultimate decision would rest with the project manager regarding the issue at hand.

On the other hand, if the project manager and the team member are in a development-readiness situation such that the project manager is more of a coach or partner with the team member in making decisions regarding the team member's tasks, then it is likely that both of them own the project problem and should work out the solution together.

Finally, if the development state of the team member is such that he is ready to make most of the decisions regarding his project responsibilities by himself, then the team member would own the problem and the project manager would likely simply be there to assist if needed.

Two warnings are appropriate: (1) It is very possible that neither the project manager nor the project team member owns the project, but rather it is owned by a line manager or other stakeholder. Do not try to take ownership of a problem you shouldn't own! Instead, help open the way for the rightful owner to manage the conflict. (2) Don't try to resolve conflicts for other people when they own the conflict; instead, you might better just facilitate or coach the process of resolution, as needed. For example, if you step in and try to resolve a conflict between two of your employees, you could weaken their self-esteem and strength to resolve their future conflicts, and it can cause them to expect you to resolve those future conflicts; conflicts they might be able to resolve themselves. Build your employees, don't weaken them. Show them the way, but don't do for them what they could do themselves with your guidance.

How Important Is the Conflict? Very early in managing a conflict we should determine just how important the conflict is. Some conflicts are extremely important and, if not properly managed, can cause serious damage to our objectives and possibly to some of the people involved. Other conflicts might only be moderately important with respect to their impact on our work and people, but these still need attention and proper management or they could grow and lead to more serious consequences. Finally, some conflicts are really of very little or no important consequence and could possibly even be ignored. Still, we must not just brush off a source of conflict as being unimportant, without first carefully listening to the people involved and evaluating the consequences of any actions taken, or not taken, to manage the conflict. *One must always remember that failing to deal with any conflict early could cause the conflict to become big!*

How does one decide the importance of a conflict and what conflict management steps should be taken? Well, this is difficult to tell in many cases without sincere communication and without honest evaluation of our work objectives, the organizational mission and strategy, and the needs and objectives of the people involved—listening and this type of evaluation are almost always needed. But there are some generalized approaches that can help.

Conflict Management Approaches

Conflict management approaches or styles can be categorized into six different categories: *preventing, avoiding, competing, accommodating, compromising, and collaborating or synergizing*. For our purposes here, *prevention* and *avoiding* are not the same approach, as will become clear in the discussion below. As you study each of these styles, ask yourself the question, "Under what conditions would I want to use this style and how would I do it?" Also, ask yourself the question, "What principles are involved with my decision as to which style to use and how I do it?"

Preventing. As explained at the beginning of this paper, you do not want to prevent all conflict on a project. In the first place it would be undesirable, and in the second place to do so would be impossible. However, it is usually unwise to cause unnecessary conflict purposely, unless you have a very carefully thought out and fully justified reason for doing so—you will usually get enough conflict without causing it!

For example, if you are a project manager, during project initiation the failure on your part to clarify the project objectives will almost certainly lead to unnecessary and wasteful conflict—do not let this happen if you can possibly avoid it. For another example, your lack of planning for a team meeting will likely lead to unnecessary conflict—again this should not happen. The bottom line here is that you should be preventing unnecessary conflict by practicing good management procedures and by following good management principles! There is a good reason for the old saying, "An ounce of prevention is worth a pound of cure"—because it is true.

Avoiding. Avoiding conflict, different from preventing conflict, is an approach in which you deal with a conflict by either indefinitely delaying action, or at least delaying action until some specific time in the future, or alternatively you choose to take no action at all. This does *not* mean that you ignore the conflict. In fact you might listen to the parties involved, ask clarifying questions, and give feedback showing that you understand the situation, but you take no action at this time. You also should take care to monitor the results of your choice to avoid the conflict, as it is possible for avoided conflict to grow to where it can no longer be avoided.

Some of the reasons we might choose to avoid conflict are:

→ The conflict is simply not worth the time and effort necessary to address it properly—the effort cannot be justified.

→ We feel overpowered and essentially helpless to do anything about the conflict.

→ Values are the source of conflict—as discussed above we cannot change a person's values.

→ We feel threatened.

→ We feel we cannot win—the situation is either lose/win or lose/lose.

→ We might just feel a need to delay action, for any of a number of possible reasons.

→ Our experience tells us that this source of conflict is best avoided.

→ We have too many more important "battles" to address, and we must "choose our battles" in order to win the war.

Avoiding conflict can be a very good way to deal with some conflicts initially—sometimes we just need a little time for conflicts to simmer so we can gather our wits and develop a better approach to the particular conflict at hand. Again, this is not the same as ignoring the conflict, which is usually not a good approach.

One very good reason to avoid a conflict is if we feel emotional about it. We frequently lose our best sense of judgment when we are emotional—we tend to see only our side and not that of other people in the proper perspective. We can also make serious and costly mistakes when we are filled with emotions. All of this is especially true if your emotions have risen to the point of your being out of control—at that point you are wrong, you need to stop, and avoid the conflict until you can gather your wits and act in a saner manner.

Finally, whenever avoiding conflict, we must assess the risks involved—what will be the effect on you personally, on other people, on the project, on reputations, on your integrity and that of others, and so on. Be careful. Avoiding conflict can be a valuable tool, or it can be one that is easily misused and dangerous.

Competing. Flannes and Levin (2001) state that competing "is often driven by a hunger for power where individual concerns and goals are pursued at the expense of others." *Competing* frequently happens when a decision is very important and when one or more of the following conditions are present: (1) there is a state of high time-pressure, like emergencies, (2) there might be a one-time only interaction with the people involved, (3) when you feel yours is the only side that is right, (4) when you cannot afford to lose, or (5) when there is a power struggle occurring.

We hope it is obvious to the reader that this approach should be used with great caution and not be used very often—only when absolutely necessary—because this is clearly a win/lose approach—I win and you lose. By its very nature such an approach can cause hard feelings, leave people de-motivated and feeling depressed or angry, cause people to seek their own interests rather than the good of the team and the organization, and cause an environment of distrust and suspicion. This style can also breed additional conflicts for the future.

The benefits of the competing style are (1) that decisions can sometimes be made much more quickly, (2) that you might force feelings to come out into the open, (3) that your views have been clarified, (4) that you get your way, and (5) that the "right" decision is made—meaning the one you feel is right. You have to ask yourself the question "Is this approach really worth the consequences?" If the answer is "yes," then you might want to use it. The problem is that people often answer "yes," when they should not. Good interpersonal relations with other people cannot last long on a win/lose diet.

To avoid the overuse of this style, you might want to find someone who is both very experienced and neutral, and have that person either advise you in your course of action or actually mediate the management of the conflict. When you can, find a better approach over that of competing.

Accommodating. Accommodation is an approach in which we subordinate our own desires to those of another person or persons in order to satisfy the other person.

There can be a number of reasons to accommodate someone else.

→ Sometimes an issue is so important to the other person that it can be worth the sacrifice to us individually to let the other person have his way, in order to maintain a positive relationship with the other person. This can be particularly true if the decision to accommodate really has very little or no negative impact on the outcome of our needs, but instead has a positive effect on interpersonal relations and teamwork.

→ Sometimes you are in a subordinate position and political sense tells you that to accommodate is better than to get your own way.

→ Sometimes you have less information or knowledge than another person and it is better to let the other person's judgment prevail over your own.

→ Sometimes it is simply not worth the effort or the pain to try to get your own way.

→ Sometimes you have to say to yourself "I have to choose my battles to win, and this is not one I choose." Someone once said that on a team, if you are winning more "battles" than the other people on the team, you might need to back off and stop dominating. This can be very good advice.

On the other hand there are a number of cautions that must be considered when accommodating. For example, if you accommodate too much, you might be perceived to be a weak manager, leading to a *ship without a rudder* syndrome. When you accommodate someone, other people could possibly want you to accommodate them, too.

You have to be careful that you don't continue to accommodate to the point that your own emotional bank account becomes depleted to the point that you are placed on the defensive and begin to lash out irrationally or in desperation to save your own feeling of self worth. The authors recommend that the reader read *The Seven Habits of Highly Effective People*, by Stephen Covey, to learn many valuable insights into emotional bank accounts. This concept is well worth the time for everyone to learn about.

Because accommodation is generally lose/win, it is usually not as good an approach as win/win approaches. Used carefully, accommodation can be very useful for the reasons discussed above; abused it will be frustrating and debilitating. We talk next about two alternative win/win approaches to conflict management.

Compromising. Compromising is accomplished when both parties in a conflict get some of what they want and don't get some of what they want. We might compromise when we are under severe time constraints and cannot reach a better solution in the time required. We might compromise in order to prevent a lose/lose, a lose/win, or a win/lose solution. We might compromise when there is no way for both parties to fully win. Or we might compromise because again, "This is not one of the battles I choose to win."

Compromising tends to be short-term, because we really have not fully satisfied any of the parties involved. Thus, conflicts may grow out of the long-term dissatisfaction that remains. So, we must not neglect consideration of the long-term consequences of compromising. Compromising can cause people over time to be perceived as weak, and again we run the risk of creating the impression that we are not able to make the needed decisions for the job—which could be true. There may be other long-term consequences affecting other people, projects, and organizations when we compromise. In decisions about compromising, we need to consider what is best for both the organization as a whole and the individuals involved.

One important aspect to consider when compromising is that you must never compromise your integrity, no matter what. Without integrity, little else really matters.

Compromising is a very low form of win/win. We both get some of what we want. The problem is that we both have to give up some what we want, too. Therefore, the risks are high that we might both end up unhappy or unsatisfied. So, in general we compromise when we have to, but there is still a better way: collaborating or synergizing.

Collaborating or Synergizing. When we collaborate or synergize, we get something greater than the sum of the parts. For example, when you and I have a conflict, both of us probably have some good aspects or perspectives which we don't want to lose in managing the conflict. The difficulty is in coming to a point where either the resolution of the conflict or the management of the conflict is equal to or better than our expectations. This can happen when we synergize.

When we synergize, we take each other's strengths and each other's differences, and we look for a new way to put these strengths together so that we get something better than we wanted. This doesn't mean that we get *what* we wanted, it means we get *something better* than what we wanted—we synergize. This is not compromise; this is better! This is true win/win. Covey calls this process one of looking for the third alternative; it is not my way or your way, it is a better way.

Collaboration or synergizing is usually the best way to manage conflict, if we can reasonably achieve this within the constraints of the situation. We want to collaborate or synergize when we

→ Value our long-term relationship—which should be the case in most situations,

→ Share similar goals or objectives,

→ Have joint ownership of a project, situation, problem, or need,

→ Have time available,

→ Both have something important to add to the solution,

→ Want a better product, and

→ Want both sides to benefit.

The authors hope that the reader realizes the tremendous potential in collaboration. In most situations we encounter, we have opportunities for collaboration. In the few where we cannot collaborate, we need to use great caution and carefully assess the risks and benefits of the less-desirable alternatives listed above.

Why Not Try to Minimize Conflict?

Above, we have dealt with conflict as a given, and yet there are ways to help minimize unhealthy types of conflict before they grow. Here are a few thoughts.

→ Stop and think about a situation before you react. This allows you to mentally take a *time-out* and get a hold of your emotions before you react and do something you either regret doing or cannot take back after being said.

→ Establish trust. Establishing trust between you and your coworkers will help promote a healthy relationship and minimize the possibility of conflict.

→ Try to comprehend the underlying motives of the conflict. Many times things are not always what they seem at first glance. Further investigation may help the situation to come to resolution.

→ Keep meetings under control. By doing this, all parties involved are less likely to get carried away and say or do something that may jeopardize them, the management, or other team members or employees.

→ Listen to all involved parties. It is easy as human beings to show or express our prejudices and/or stereotypes; however, when we listen to all involved parties, the playing field or conflict discussion table is fair and level.

→ Develop a give-and-take attitude when dealing with conflict. A lot of times, employees or managers end up selling their own self short in a conflict when a little give-and-take could have made a world of difference, avoiding grief and despair.

→ Value the differences in people. Much better decisions and communications occur when we truly want and sincerely try to understand what others have to offer. Be humble enough to be teachable by the vast reserves in others.

→ Try hard not to jump to conclusions. Sometimes it's best to "go slow."

→ When you sense conflict, try looking in the mirror first—you might be the cause. The saying, "Physician, heal thyself," may well apply.

→ Gracefully help others to know and understand your viewpoints and opinions. By doing this, you may see fewer conflicts because the other employees already know how you feel. When you do this, always remember the Golden Rule to do unto others as you would have them do unto you.

→ Clarify expectations! It is amazing how much conflict arises from a lack of understanding of expectations. The clearer you can make expectations, the easier it will be for you and for other people to measure results and meet those expectations.

→ Reward good performance openly and correct poor performance privately.

→ Respect confidences! If you don't, you are highly likely to be the center and cause of conflict, and you will be wrong.

→ Don't play favorites. Be fair. Again, follow the Golden Rule.

→ Admit when you are wrong. Nobody likes to admit when they are in error, but conflicts may subside when people are open, honest, and responsible for their actions and decisions.

Some Final Comments Regarding Conflict Management

Leadership has a great deal to do with conflict management. Great leaders will use all of the conflict management approaches we have discussed in this chapter, but they have the good sense to use the various approaches in a powerfully balanced way. The result is that they are able to have high deposits in other people's emotional bank accounts and at the same time accomplish great feats while overcoming tremendous challenges.

References

Steven W. Flannes and Ginger Levin, *People Skills for Project Managers*. Management Concepts, Vienna, VA, 2001.

Stephen Covey. *The Seven Habits of Highly Effective People*. Free Press, New York, NY. 2004.

240c Smart Manufacturing Plants and Smart Management

Jimmy L. Humphrey, J. L. Humphrey & Associates, 3605 Needles Dr., Austin, TX 78746, J. Christopher Lewis, Technology R&D, ConocoPhillips, 166 PDC Bartlesville Technology Center, Highway 60 & 183, Bartlesville, OK 74004, and A. Frank Seibert, Separations Research Program, University of Texas at Austin, 10100 Burnet Road, Bldg 133/CEER, Austin, TX 78758.

Smart Manufacturing Plants prevent problems before they occur. Smart people are the most important ingredient in Smart Plants. Examples of problems Smart Plants prevent include excessive energy use, off-spec products, unplanned shutdowns, and unwanted safety and environmental events. Accurate real time plant data, obtained from key sensors and measurements, are the foundation of the Smart Plant. These data establish current performance and are used with models to PREDICT future plant events so PROACTIVE actions can be taken to PREVENT problems before they occur. A three-step process will be presented to prevent plant problems before they occur.

240d Abet Accreditation as An Exercise in Cooperation

Richard L. Long, New Mexico State University, Chemical Engineering Department, P.O. Box 30001, MSC 3805, Las Cruces, NM 88003-8001

The ABET 2000 accreditation requirements constitute a cascade control loop process that requires cooperation among the stakeholders of a program to achieve success. The "program objectives loop" becomes the "master loop" for the "program outcomes loop" as the "slave loop". The ABET 2000 requirements force the program to demonstrate that it has identified deviations from desired objectives and/or outcomes, measured those deviations, and acted to close the respective loops based on the deviations observed. Stakeholders must provide input to the process of defining the "objectives" and "outcomes" desired. The process requires a much broader and deeper participation and cooperation of stakeholders than the older accreditation requirements.

VitalSmarts™

The Concours Group

CONTENTS

AN INTRODUCTION TO *SILENCE FAILS*

The Five Crucial Conversations for Flawless Execution

Ninety-one percent of all large-scale corporate initiatives fail—initiatives that have been designed to save a company from impending disaster, increase organizational competitiveness, or just accomplish strategic objectives. They consume unplanned resources and miss schedules, they leave behind the stench of failure, and they cost leaders careers.

Yet many organizations continue along the same precarious path year after year—accommodating behaviors that contribute to these costly mistakes.

How can we avoid repeating history? Why do projects that demand so much of a company's top leaders' time and attention collapse? Do people just not know how to plan? Do they fail to involve experts? Are they hopelessly incapable of making accurate projections? No. History repeats itself when teams and organizations—made up of people—fail to speak up well and be heard. Unfortunately, a deadly form of corporate silence lies at the root of all failed projects. Like it or not, project problems are people problems.

In the following pages, you will find some startling results that will change the way you look at your next initiative, program, or project—and demonstrate that you can get better. This groundbreaking research, *Silence Fails*, outlines findings that are simultaneously daunting and hopeful. For the first time, *Silence Fails* accurately contrasts the causes and costs of failed initiatives are accurately contrasted with the impressive returns of investing in skills that can help anyone navigate the pitfalls of project execution

How Can You Improve

For more than twenty-five years, the founders of VitalSmarts have researched methods for bringing about systematic and lasting change. Based on this ongoing research, VitalSmarts has helped thousands of organizations, including more than three hundred of the Fortune 500, realize quick, hard-hitting results.

Based on this extensive research, here are some tips for getting started:

1. Take the *Silence Fails Assessment* found at the end of this report. This assessment highlights the points at which your initiative, program, or project is likely to fail.

2. Contact a VitalSmarts Representative at 1-800-449-5989 to request a free customized team version of the *Silence Fails Assessment*. This assessment is Web-based, allowing your entire project team to rate a project and view the results.

3. Distribute this report to help others become aware of the problems it reveals.

4. Once you've increased awareness and appreciation for the issues, it's time to move to action. A VitalSmarts Representative can help you develop an action plan that will meet your specific needs. Call today to find out how you can begin to change your organization.

Other tips for change are available in the conclusions section of this report and at www.silencefails.com.

THE *SILENCE FAILS* STUDY

The Five Crucial Conversations for Flawless Execution

The Silent Crisis

This year, leaders across the world will make an awful bet. They will bet a substantial portion of their organizations' budgets on ventures that have less chance of paying out than a red or black bet on a roulette wheel. And yet, as in every year for many decades, leaders will continue to put the money down.

Their preferred hedge against failure? Formal systems. Over the past twenty years, project professionals and management experts have focused on improving the formal systems related to program governance, project management, and project-related technologies.

KEY FINDINGS

Breaking a code of silence on five astoundingly common yet largely undiscussed and ignored problems will contribute significantly to project success.

While these new approaches have substantially improved results, today two-thirds or more of initiatives, programs, or projects still miss their mark. Something is clearly missing. Beginning in late 2005, VitalSmarts and The Concours Group partnered to dive below the formal systems to identify what is missing when project execution fails.

This study, *Silence Fails: The Five Crucial Conversations for Flawless Execution,* demonstrates that project leaders can substantially improve their organization's ability to execute on high-stakes projects and initiatives by breaking a code of silence on five astoundingly common yet largely undiscussed and ignored problems that contribute significantly to almost all project failures.

Situation Analysis

Senior executives frequently bet their companies on high-stakes efforts like major product releases, strategic IT projects, organizational restructurings, fast-paced downsizings, or aggressive quality initiatives. And these bets rarely pay off as anticipated. With estimated failure rates ranging from 72 to 91 percent[1], companies' collective inability to execute on major projects costs hundreds of billions of dollars a year. For example, it's estimated that of the $255 billion spent per year on IT projects in the U.S., more than a quarter is burnt up in failures and cost overruns.[2]

The failure of the cutting-edge baggage handling system at the Denver International Airport provides a dramatic example of this inability to execute. With 26 miles of conveyors, the system promised to carry bags more quickly and reliably from passengers to planes and back again. Instead, the breakthrough concept racked up hundreds of millions of dollars in cost overruns, delayed the airport's opening by over a year, and drained 10 years of additional resources in various salvage attempts before leaders admitted failure.[3]

In addition to sapping organizational performance, these project failures cost careers. Now more than ever, CEOs are under pressure to either get results or get lost. In 2005, CEO turnover doubled from the year before.[4] In just the past five years, close to two-thirds of all major companies have replaced their CEOs. CIOs are similarly vulnerable, with a quarter losing their jobs each year.[5] Studies suggest that an inability to deliver on critical projects is a primary reason for this alarming rate of dismissals. Their shortcomings were less about strategy and decision making than about their ability to execute these plans.[6]

Five Crucial Conversations

Silence Fails suggests that leaders can significantly increase their organization's ability to execute on high-stakes projects and initiatives by focusing on the five crucial conversations outlined below. The study also found that increased effectiveness at any one of the five crucial conversations that address these problems improves project performance by 50 percent to 70 percent or more.

The five crucial conversations identified by the study are the most prevalent and most costly barriers to project success. Each brings with it myriad misses and overruns that plague projects, teams, and organizations. *Silence Fails* addresses the cost, culture, and long-term dynamics of each. They are:

1. **Fact-Free Planning.** A project is set up to fail when deadlines or resource limits are set with no consideration for reality.

2. **AWOL Sponsors.** A sponsor doesn't provide leadership, political clout, time, or energy to see a project through to completion.

3. **Skirting.** People work around the priority-setting process.

4. **Project Chicken.** Team leaders and members don't admit when there are problems with a project but instead wait for someone else to speak up.

5. **Team Failures.** Team members are unwilling or unable to support the project.

KEY FINDINGS

When these problems are not openly and skillfully discussed, projects of all sizes fail miserably.

This study confirms that some projects are plagued by more of these problems than others—likely due to high levels of interdependence among levels and functions. And when these problems are not openly and skillfully discussed, projects of all sizes fail miserably. Those who effectively discuss these issues have a positive impact on the objectives of their projects.

Key Findings

Silence Fails demonstrates that these five problems are not a project death sentence, but they significantly impact the results a team or organization may be trying to achieve. Findings include:

- 90 percent of project leaders regularly experience one or more of these problems with nearly 40 percent of their projects.

- 91 percent report that these problems are difficult if not impossible to resolve in their current organization.

- 61 percent attempt to speak up about the issue, yet only 14 percent feel they speak up in a skillful way that has a lasting effect on the problem.
- Up to 88 percent claim that these issues continue over the life of their project and beyond.
- Between 82 percent and 87 percent of projects report significant negative impact when they fail to address budget, scope, and schedule issues.
- Team morale suffered significant setbacks among 73 percent of projects studied. 43 percent of projects continued to suffer from a long list of problems even after the project ended.

Silence Fails demonstrates that whether these five problems can be discussed determines whether or not they will derail an initiative. The data show that when people discuss these problems in effective ways, projects come back on course. And when one or more of these five problems is not confronted—or not confronted well—it festers, sets off workarounds, and produces politics. In the end, project failure is almost inevitable.

KEY FINDINGS

When one or more of these problems is not confronted–or not confronted well–it festers, sets off workarounds, and produces politics.

The Good News: Speaking Up Makes a Difference

While the data show that fewer than one in five project leaders effectively engage in the crucial conversations needed to solve these five problems, the good news is that one in five *does.* Much can be learned by studying their successes and whether these skilled individuals produce different results than their peers. Research on this contingent of crucial conversations successes helps answer questions like:

- Does it affect project success when project leaders speak up effectively?
- Can others be taught to speak up more skillfully with similar results?

The resounding answer to each is "yes."[7] In each of the five key problem areas, the study found a clear difference between speaking up and speaking up well. Across the problem areas, about half of leaders make some attempt to speak up. But most are ineffective. Some speak up but water down their concerns, so the issues are never fully aired. Some speak up but do so in a way that provokes defensiveness from others. And a handful—about one in eight—are able to share their full concerns and, by the end of the conversation, feel their views are understood.

The research also showed that while the skill of the initiator is a key ingredient in ensuring these crucial conversations are held well, the receptiveness of the other party is similarly important. Project leaders who are adept at holding crucial conversations are far more effective in dealing with prickly and defensive executives. But their skill doesn't guarantee their concerns will be addressed. Unless and until leaders take extraordinary measures to ensure their environment is conducive to holding crucial conversations, a significant number of issues will remain unaddressed, invisible, and fatal.

However, when a skillful leader creates even a moderately safe environment:

- The likelihood of a project failing to meet budget is reduced by almost half.
- When project leaders effectively step up to these crucial conversations, hitting schedule is at least 40 percent more likely, quality improves almost 60 percent, and the likelihood a project will end with strong morale and intact stakeholder relationships is almost 70 percent greater.

KEY FINDINGS

Unless and until leaders take measures to ensure their environment is conducive to holding crucial conversations, a significant number of these issues will remain unaddressed, invisible, and fatal.

Next Steps for Senior Leaders

Although *Silence Fails* focuses on important findings that can predict and explain failure, the most important implication of the study is the potential leaders have to influence success.

Senior leaders can predict and prevent the failure of high-stakes business initiatives by creating a culture where the five conversations are held quickly and effectively. *Silence Fails* also provides insights and recommendations on how senior leaders can develop a business case for change, measure behaviors, invest in skills, hold senior management accountable, and make heroes of early adopters.

Organization of This Report

This report contains the results of the *Silence Fails* study. Each of the five crucial conversations for flawless execution is explored, including the following:

- A composite case study, The Last Roundup, that describes the issues the *Silence Fails* research team documented in actual projects.
- Key findings as they relate to each specific crucial conversation, including the extent of each problem, the cost of not speaking up, and the effect of speaking up.
- Actual comments from study participants that further reflect the environment and effects of these crucial conversations when they go unaddressed.
- A conclusion that identifies best practices for leaders who want to help their organizations improve their ability to step up to and hold these five crucial conversations to improve project success.

Section Notes

1 "CHAOS Chronicles," Standish Group, 2004 and "The Strategy Focused Organization," Kaplan and Norton, 2001
2 "CHAOS Chronicles," Standish Group, 2004
3 "Dependable Software by Design," Scientific American, June 2006, p. 69
4 Challenger, Grey, and Christmas, cited in USA Today, 10 April 2006, p. 2B
5 "Work Management: Why Can't Information Managers Manage?" John J. Lucas, PMI Seminar/Symposium. Proceedings, 1995, p. 304 310.
6 Study by LeadershipIQ.com
7 Based on over 25 years of research, VitalSmarts continue to conduct field studies that demonstrate significant, sustainable results brought about through organizational change initiatives. These efforts focus on behavior modification as opposed to system design and strategy development. See www.vitalsmarts.com for case studies and additional information.

ABOUT THIS CASE

This case follows a composite character, Bob Kelley, through a major organizationwide initiative, and describes issues the *Silence Fails* research team documented in actual projects.

Throughout the remainder of this study, pertinent Silence Fails findings will be applied to the this case study.

CASE STUDY:
THE LAST ROUNDUP

A project leader in a major business services company, Bob Kelley led an effort codenamed "Roundup" to in-source the company's billing system. The history of the billing system was complex:

- The billing system was the lifeblood of the company—responsible for drawing hundreds of millions of dollars into the company on a monthly basis by issuing timely and accurate bills to more than 20 million customers.
- For twenty years the billing system had been handled by a vendor.
- For three years Kelley led a sometimes unruly and loose confederation of more than 1,000 professionals drawn from various functions in his company, as well as contractors to change over to a company-owned system.

Roundup was a particularly tough assignment given that Kelley was the line manager of only a few of these people, and his mission depended on the unflagging support of *all* of them.

The stakes couldn't have been higher. And Roundup's status could not have been worse. By the time we met Kelley, the project was already 100 percent overdue and 300 percent over the original budget—an investment of well over $250 million to that point with no end in sight. Kelley believed, however, there was a light at the end of the tunnel. The first cutover date for a small contingent of customers was two weeks away. As he walked out of a "release review," he believed all was in order and the chance of success was high.

One month later, Kelley was handed his head. The release was a disaster. The vendor managing the legacy system failed to transfer data when required. Once they did, the new bills were plagued with errors, and the system suffered from multiple "Severity 1" glitches—causing it to crash uncontrollably during the billing cycle. When he was dismissed, Kelley was escorted to his desk by a security guard who then walked him out of the building.

Diagnosis

Kelley's demise was the direct result of failure to effectively address the five issues identified in *Silence Fails*. Each of the five key problems emerged at one time or another during the course of the initiative.

Conclusion

The complete undoing of Roundup was not guaranteed by the existence of these problems but by how they were handled. Most were never fully discussed with the people who needed to hear them. Others were raised, but then muffled by the very executives who should have listened and taken action.

Fact-free planning reflects bad planning behaviors at every level.

Extent and Costs

Of the 85 percent of study participants who reported the prevalence of fact-free planning in their projects:

- 39% of their projects were affected
- 91% said the issue was difficult or impossible to confront or resolve
- 61% spoke up at all
- 14% spoke up skillfully and had some effect on the problem
- 19% were able to solve the problem

88% reported that the problem persisted*, causing operational impact as follows:

- 82% of projects exceeded their budget
- 82% missed deadlines
- 82% had missing or wrong functionality or quality problems
- Team morale was damaged on 73% of projects
- 43% had a long list of problems to be resolved after the project ended

* Continued for the life of the project and beyond.

CRUCIAL PROBLEM #1:

FACT-FREE PLANNING

A project is set up to fail when deadlines or resource limits are set with no consideration for reality.

Fact-free planning describes ways leaders set project parameters without involving—or by falsely involving—project leaders. Project leaders are often given sets of deliverables, budgets, and timelines with no input or opportunity to determine what the project will really take. At other times they are asked to calculate the resources and time needed to achieve a scope of work, but then their estimates are ignored. In either case, the result is a set of timelines, budgets, and deliverables that are unrealistic from the beginning.

Very often, unrealistic constraints get placed on projects. An executive or customer will make commitments to another stakeholder without the project team's consent and will then present the finished plan as a done deal. Over time, teams that are subjected to these mindless demands begin to pad budgets and timelines—anticipating that their estimates will be ignored.

Fact-free planning reflects bad planning behaviors at every level. When project leaders realize these practices are taking place, they must be willing and able to call the bluff. If they avoid this crucial conversation and either commit to something they know can't happen or fake their way to success, they set themselves and their projects up to fail.

Similarly, executives who avoid discussing their doubts about an estimate's validity or the team's competence instead use their power in a way that generates *political* rather than *valid* agreements. Then when failures follow, their doubts about the team are confirmed, and they feel justified in using more fact-free planning to re-establish their sense of control.

The only way out of this vicious cycle is for project leaders and executives to candidly and effectively express their suspicions, doubts, and data. While these crucial conversations aren't easy, they are the only path to rational commitments.

How common are these problems? Eighty-five percent of project leaders are faced with "fact-free planning." Key drivers include:

Indicators of Fact-Free Planning

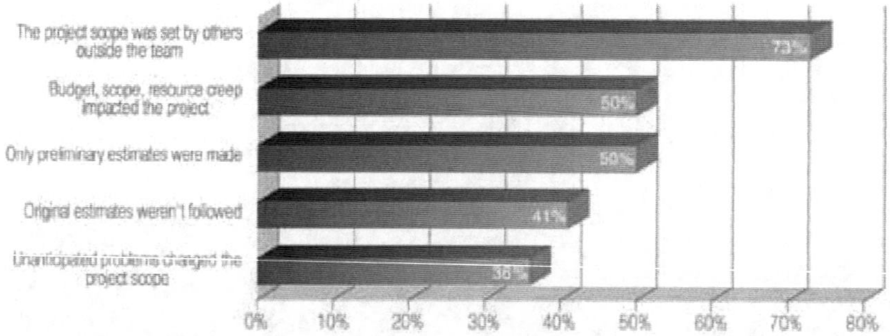

Impact of Speaking Up

The 19% who skillfully addressed this problem** improved quality and functionality by 69% while reducing:

- Cost overruns by 29%
- Schedule delays by 17%
- Inappropriate shortcuts by 47%
- Persistent problems* by 58%

* Continued for the life of the project and beyond.
** Compared to those who didn't address the issue (these differences are all statistically significant, $p < .05$)

Application
THE ROUNDUP CASE

When Kelley took over Roundup, he was given one year to complete the insourcing and a budget of $45 million. Ideally, this schedule and budget would have been the result of a careful planning process that involved Kelly, but that's not the way it happened. Instead, Kelly's executive leaders were playing a common game in project execution—they fished for the best deal they could get. They set a stretch goal and hoped the performance pressure it created would motivate him to achieve it. Kelley was a good soldier and tried to make it work, but he knew from the beginning that it couldn't.

Just how this plays out with project teams is better illustrated by these key findings:

- When project leaders face these problems, they know the stakes are high, and 61 percent make some effort to raise their concerns—often in an understated way.
- More than nine out of ten project leaders describe this type of conversation as between "difficult" and "impossible" to resolve.
- In fact, only 19 percent are able have the conversation in a way that gets results. When project leaders are unable to succeed at this conversation, their projects turn into slow-motion train wrecks.

When the conversation fails, the impending failure becomes widely anticipated.

- Nine out of ten project leaders share their concerns with project team members and with others across the organization.
- Team morale takes a hit in nearly three-quarters of these cases as people become aware that they're headed toward failure. At this point in the project, the unrealistic specs are rarely resolved; 88 percent of the time they remain for the life of the initiative or beyond.
- 9 percent of these projects come in over $1 million over budget.
- 70 percent of the time, people take inappropriate shortcuts, cut documentation, or skip testing in order to meet the specifications, and in the end these projects fail.
- In fact, 61 percent achieve the hat trick of project failure; they miss their deadlines, go over budget, ***and*** fail to achieve their deliverables.
- 43 percent end up with a long list of problems to resolve after the project itself has ended.

Participant Comments

"Here's how it works here. Year 1: Six Sigma is the hot thing and you get all the resources you need. Year 2: They put the squeeze on you arbitrarily cutting by X percent. Year 3: You're not meeting goals because you have no resources, and they kill it." – *Manager, Pharmaceutical Firm*

"We ask you to do a project. You come back and say, 'I need twenty-five resources and $2.5 million.' And your sponsor says, 'Okay, how about three resources, clock time, and no money?' and we say 'Sounds right.' You have to, because you can't challenge." – *Department Head, Food Services Firm*

"There's no such thing as planning for a project. There's the real date and the date you're told to do it. Part of the front end of a project should be identifying and refining your scope. What is realistic? In previous experience—it's not about what is the true scope and how many resources and how long. Senior manager A spoke with senior manager B—we're going to have this done by the end of February. We set an end date and work backward. So we cut out testing and other quality processes." – *Director, Financial Services Firm*

When the project sponsor is AWOL, the project team is stranded and exposed.

Extent and Costs

Of the 65% of study participants who reported the prevalence of AWOL sponsors in their projects:

- 33% of their total projects were affected
- 88% said the issue was difficult or impossible to confront or resolve
- 47% spoke up at all
- 11% spoke up skillfully and had some effect on the problem
- 17% were able to solve the problem

75% reported that the problem persisted*, causing operational impact as follows:

- 75% of projects exceeded their initial budget
- 85% missed deadlines
- 74% had missing or wrong functionality or quality problems
- Team morale was damaged on 69% of projects
- 31% had a list of problems to be resolved after the project ended

* Continued for the life of the project and beyond

CRUCIAL PROBLEM #2:
AWOL SPONSORS

A sponsor doesn't provide leadership, political clout, time, or energy to see a project through to completion.

Project sponsors are responsible to provide leadership and political support. And they frequently don't. When the sponsor is AWOL, the project team is stranded and exposed. They're sent off to accomplish a task and don't have the firepower needed to implement the project. For example, key leaders whose help or resources are needed to enable the project may not come through, and the sponsor who has the organizational muscle to hold them accountable fails to do so.

In many cases the sponsor who should have fought the political battles (returning calls, following up with commitments, and so on) ultimately leaves project leaders to fend for themselves. Rather than directly address the issue, they begin working around it.

Key findings include:

- 65 percent of project leaders experience problems with AWOL sponsors.
- About half of those make some attempt to bring up their concerns.
- 88 percent of study participants indicate that conversations on this subject are difficult if not impossible to have in their organizations.
- Fewer than one in five project leaders who suffer from AWOL sponsors are able to hold the crucial conversation with their sponsor in a way that solves the problem.

When the project leader fails to resolve AWOL sponsor problems, projects are destined to fail—and the costs add up.

- More than three-quarters come in substantially over budget, behind schedule, and/or below specifications.
- In fact, 11 percent come in more than a million dollars over budget.
- Team morale is damaged more than two-thirds of the time.

Participant Comments

"My project from the president was, 'We need to eliminate $50 million in cost by moving three major manufacturing programs from North America to our Asia operations.'

"My 'sponsor' was the VP of North America. I met with the North America person and he basically said, 'Over my dead body.' And this is my sponsor! I went to the president and said, 'What is up with this? How am I going to manage through this?' So he sent an e-mail to the sponsor and ordered him to help me. My sponsor is now my sworn enemy." – *Director, Manufacturing Firm*

Impact of Speaking Up

The 17% who were able to address this problem** improved quality and functionality by 50% while reducing:

- Cost overruns by 33%
- Schedule delays by 33%
- Persistent problems* by 70%
- Inappropriate shortcuts by 66%

* Continued for the life of the project and beyond.
** Compared to those who didn't address the issue (these differences are statistically significant, $p < .05$)

Application
THE ROUNDUP CASE

In Bob Kelley's case, the sponsor who should have fought the political battles required to get the old vendor to cooperate rarely returned calls, seldom followed up with commitments, and ultimately left Kelley to fend for himself.

When Kelley took on Roundup he made a fatal mistake. He was told through the grapevine that his sponsor was a bit of a deadbeat. Rather than directly address this issue, he began working around it. He called on other executives for help and enabled the problem by absolving his sponsor of responsibility.

"I was working with people I've never met in another country and basically I was trying to drive something that the CIO wanted. I tried to get people to understand that this was what we were being told to do—that it was a directive, not a choice. I finally had to talk to the CIO and explain to him, 'I need help, you need to explain why we need to do this. You need to get people's buy-in and support.' I was afraid an email would go out like 'This is what you need to do and you will do it NOW.' I don't think that would have helped the situation. Instead, he kind of backed off and the project died. But not for another year." – *Program Manager, Pharmaceutical Firm*

Skirting plagues most projects for their entire duration.

Extent and Costs

Of the 83% of study participants who reported the prevalence of skirting in their projects:

- 33% of their total projects were affected
- 79% said the issue was difficult or impossible to confront or resolve
- 51% spoke up at all
- 13% spoke up skillfully and had some effect on the problem
- 16% were able to solve the problem

81% reported that the problem persisted*, causing operational impact as follows:

- 78% of projects exceeded their budget
- 87% missed deadlines
- 80% had missing or wrong functionality or quality problems
- Team morale was damaged on 66% of projects
- 28% had a long list of problems to be resolved after the project ended

* Continued for the life of the project and beyond.

CRUCIAL PROBLEM #3:

SKIRTING

People work around the priority-setting process.

Powerful stakeholders and senior leaders often skirt the formal decision-making, planning, and prioritization processes. They need what they need, and they don't want to be burdened with practical considerations. So they work around the process. The results are often outrageous overcommitment, disappointment, and burnout. Projects get approved for which there are no resources, scope creep bloats approved projects far beyond the resources originally budgeted, and team members deliver a succession of disappointing results and suffer from battered morale.

Key findings include:

- 83 percent of project leaders contend with stakeholders and others who skirt the formal planning and execution system. The most common descriptions of skirting include:

The Many Faces of Skirting

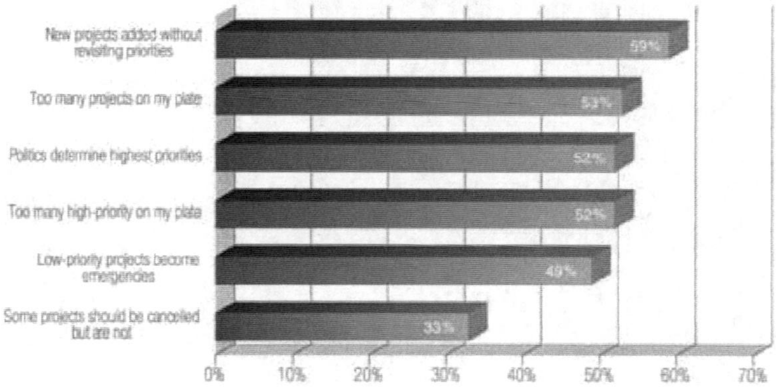

- 79 percent of project leaders describe the discussion as between "difficult" and "impossible" to resolve.
- Only 13 percent are able to skillfully raise their concerns in a way that gets heard.

With so few people able to succeed at this crucial conversation, it's no surprise that skirting plagues projects for their entire duration. Other key findings include:

- Four out of five of these projects suffer from skirting issues for the life of the project or beyond.
- 78 percent of these projects go over budget.
- 87 percent miss their deadlines.
- Two-thirds leave teams with battered morale.
- 80 percent fail to achieve their deliverables.

Impact of Speaking Up

The 17% who were able to address this problem** improved quality and functionality by 64% while reducing:

- Cost overruns by 64%
- Schedule delays by 60%
- Persistent problems* by 70%
- Inappropriate shortcuts by 66%

* Continued for the life of the project and beyond.
** Compared to those who didn't address the issue (these differences are all statistically significant, $p < .05$)

Application

THE ROUNDUP CASE

The VP of Marketing in Kelley's company repeatedly injected additional requirements into the project with no consideration for their impacts on the budget or schedule. He did so through both direct intervention and deal-cutting with sponsors and executives. Kelley and his team were left to deal with the fallout. Once again—the problem was not only that the Marketing VP skirted decision-making processes—the problem was that no one confronted his violations in a way that stopped them.

Bob Kelley and his sponsors failed to step up to the crucial conversation when repeated skirting made the scope of his project impossible to fulfill. As a result, the violations continued, and added momentum to the oncoming train wreck.

Participant Comments

"A real simple example: somebody might be walking with me to the coffee machine and they start talking about something. I have to stop and say, 'You realize this is just a conversation—that you did not just initiate a project. You cannot go back to your CIO three months from now and say, 'We've been working with them for two months now.' Talking to me on the way to the break room is not how you contract for our services." *– Senior Program Manager, Food Services Firm*

"There are people who are better at navigating through all the stakeholders. It tends to become more political. It's a lot of fine negotiating, deciding carefully who you're going to piss off and who you're not. I haven't seen an example of where we're good at balancing priorities in a traditional sense. I have just seen some better than others at being artful or persistent, or getting the right air cover to navigate through it."

"When a project manager says, 'No,' to a customer, here's what will happen: The customer will call the president of our company and say, 'You need to do this,' and he says 'Yes.' Then we get horrible scope creep because after we make agreements, people hang things on the project like a Christmas tree without adding any resources. When we fail to make schedule or budget, they don't say anything immediately, but then they badmouth us later for generally being over budget and past due." *– Project Manager, Food Services Firm*

When project participants play project chicken, the status and review process becomes a joke.

Extent and Costs

Of the 55% of study participants who reported the prevalence of project chicken:

- 31% of their total projects were affected
- 61% said the issue was difficult or impossible to confront or resolve
- 42% spoke up at all
- 13% spoke up skillfully and had some effect on the problem
- 26% were able to solve the problem

76% reported that the problem persisted*, causing operational impact as follows:

- 78% of projects exceeded their budget
- 86% missed deadlines
- 74% had missing or wrong functionality or quality problems
- Team morale was damaged on 54% of projects
- 23% had a long list of problems to be resolved after the project ended

* Continued for the life of the project and beyond.

CRUCIAL PROBLEM #4:

PROJECT CHICKEN

Team leaders and members don't admit when there are problems with a project but instead wait for someone else to speak up first.

Once a project gets underway, it can derail when various subteams or team members fail to honestly report project risks. They engage in Project Chicken.

This costly game resembles the lunatic practice of driving cars head-on as a test of nerves to see who swerves out of the way first—or who is more "chicken." The corporate version is played when project leaders fail to admit they may fall short on deliverables and need more time. Instead, they hope some other group that has problems will speak up first. Whoever speaks up first will be blamed for causing the delay, but everyone who is behind will benefit.

When project leaders play project chicken, the status and review process becomes a joke. The team loses opportunities to gracefully respond to problems by revising goals, shifting resources, reorganizing plans, and more. Instead, the project hurtles forward on a collision course with failure while everyone watches—nervous but silent.

Almost every organization reported some form of Project Chicken, and more than half of project leaders say they face it regularly. Often, the enormous pressure to give good news sets project leaders up for failure by making it difficult to discuss risks. Such pressures also make it hard for team members to admit that a single problem could pose a risk to the entire project, even though the problems may be longstanding and complex. In this situation it's also hard to admit that shortcuts could pose a risk to the entire project. Project leaders often see these trains coming for months but fail to speak up in an effective way.

Key findings include:

- Roughly one in four project leaders who face this problem are able to effectively discuss project chicken with those who appear to be playing the game.
- When project leaders face these problems they know the stakes are high, but only 42 percent make some effort to raise their concerns.
- 13 percent of project leaders describe this type of conversation as "difficult" and "impossible" to resolve.

When the conversation fails, so do 76 percent of the projects.

- 86 percent of those miss their deadlines.
- 78 percent come in over budget.
- 74 percent are missing functionality or suffer from lower-than-desired quality.

Impact of Speaking Up

The 17% who were able to address this problem** improved quality and functionality by 56% while reducing:

- Cost overruns by 55%
- Schedule delays by 69%
- Persistent problems* by 70%
- Inappropriate shortcuts by 66%

* Continued for the life of the project and beyond.
** Compared to those who didn't address the issue (these differences are all statistically significant, $p < .05$)

Application
THE ROUNDUP CASE

The enormous pressure to give good news set Kelley up for failure by making it difficult to discuss risks. It made it hard to admit that a single problem vendor could pose a risk to the entire project, even though the problems with the vendor were longstanding and numerous. It made it hard to admit that shortcuts during testing could pose a risk to the entire project, even though testing had been seriously limited to meet the aggressive release schedule.

The problem was that none of these risks were addressed either in Kelley's final project review or in prior sessions. Project participants had seen these problems coming for months, but no one had spoken up in an effective way.

Participant Comments

"There are plenty of cases where someone has known something was wrong but didn't speak up soon enough. So yeah, if you have project updates and you have good, good, good, good, good, then bad at the end of the day. Then it's too late to do anything." – *Project Manager, Construction Firm*

"Oh, yeah. We have information about project success. We claim we hit schedule every time. But it's a lie. So the statistics looks good—but what we never admit is how many times we've moved the schedule. That's the real story—but we don't talk honestly about when we won't hit schedule and why. I can't think of a single project in the past two and a half years that was finished on the date initially planned." – *VP Information Services, Manufacturing Firm*

"This was not an IT person, and she was leading a project that required a lot of IT. The people below her weren't giving her all the information that they should have, and she wasn't able to connect the dots. They knew they were in trouble, but the project was at a high visibility. They just kept working hard—like they'd eventually get there.

"I took the project from green to red overnight, two weeks before it was supposed to go live. I had to involve the second and third highest persons in our company, and obviously that didn't go over well. The person who was the project lead, I felt bad for her." – *Vice President, Food Services Firm*

Too often, team members will ignore team deficiencies and work around the problem.

Extent and Costs

Of the 80% of study participants who reported the prevalence of team failures in their projects:

- 30% of their total projects were affected
- 76% said the issue was difficult or impossible to confront or resolve
- 49% spoke up at all
- 14% spoke up skillfully and had some effect on the problem
- 24% were able to solve the problem

80% reported that the problem persisted*, causing operational impact as follows:

- 73% of projects exceeded their budget
- 82% missed deadlines
- 77% had missing or wrong functionality or quality problems
- Team morale was damaged on 69% of projects
- 24% had a long list of problems to be resolved after the project ended

* Continued for the life of the project and beyond.

CRUCIAL PROBLEM #5:
TEAM FAILURES

Team members are unwilling or unable to support the project.

Project leaders generally have a common attribute: they have far more responsibility than they have authority. They "lead" large cross-functional groups with very few direct reports. Often, all have other masters—and it shows.

Project leaders report being hobbled by team members who don't show up to meetings, fail to meet schedules, or lack the competence to meet ambitious goals. Often these leaders have no say in selecting or replacing these nonperformers and feel powerless to coach them. Instead, they ignore their deficiencies and work around the problem.

For example, someone assigned to the team may be viewed by others as incompetent. Rather than addressing such concerns openly, additional resources are brought in to take over the most critical tasks. Not only is this a costly duplication of effort, it creates an environment of resentment and mistrust between the now-competing individuals and among other team members as they are dragged into this jockeying.

A second source of team struggle comes when project leaders must negotiate with functional managers to staff their projects. These functional managers often have many other priorities and are unresponsive to the project leader's needs. Once again, if project leaders can't hold effective crucial conversations with functional managers who fail to deliver, then project success is compromised.

Key findings include:

- When project leaders fail to effectively address performance problems in their teams, four out of five projects suffer from budget, schedule, and/or quality problems.
- Of the 80 percent of leaders who reported team member failures, their observations about common causes include:

Where Team Members Let Each Other Down

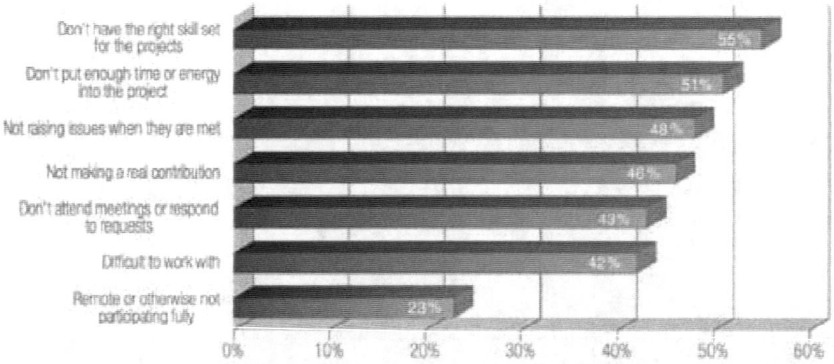

Impact of Speaking Up

The 17% who were able to address this problem** improved quality and functionality by 64% while reducing:

- Cost overruns by 64%
- Schedule delays by 60%
- Persistent problems* by 70%
- Inappropriate shortcuts by 66%

* Continued for the life of the project and beyond.
** Compared to those who didn't address the issue
(these differences are all statistically significant, $p < .05$)

Application

THE ROUNDUP CASE

From the beginning, Kelley suffered from the common lot of all project leaders—he had far more responsibility than he had authority. He "led" a group of 1,000+ professionals--only seven of whom reported to him. All had other masters—and it showed.

Kelley, for example, was assigned a senior systems architect whom Kelley thought was incompetent. Rather than addressing his concerns openly, he brought in a second architect and made sure this person received the most critical tasks. Not only was this a costly duplication of effort, it created an environment of resentment and mistrust between the two architects and among other team members they dragged into their jockeying.

- 76 percent of those describe the discussion as between "difficult" and "impossible" to resolve.
- As a result, only 14 percent are able to skillfully raise their concerns in a way that gets heard.

With so few people able to succeed at this crucial conversation, it's no surprise that team failures begin in the earliest stages of projects and persist for their entire duration.

- 80 percent these projects suffer from team failures for the life of the project or beyond.
- Up to 82 percent of these projects go over budget, miss deadlines, and suffer from quality problems.
- More than two-thirds leave teams with battered morale.

Participant Comments

"The functional managers? You've got to hound them. I just had this one project. I kept reaching out to the manager of the resource group I needed. First I tried to invite them to the kick-off meeting so they'd see the scope of their involvement. Then I sent them private e-mails. I was getting blown off for weeks. I stop by their office, and if I ever brought up how they've been blowing me off, they really didn't like it. And escalate? The most my manager would do was send them an e-mail. No one really confronted the problem. That really stinks." – *Project Manager, Food Services Firm*

"If your project isn't hot you won't get the resources they committed to you. They don't take you seriously until you have a crisis. When it becomes a crisis then we will address it. Now someone can step up and become a hero. Say the regulators are coming next week! Now you get the resources, the money, and away you go." – *Project Manager, Financial Services Firm*

"Once they assign a bad resource you can only escalate to their manager. You just accept what they give you. Or just give their work to someone else. If that person has a track record, then obviously the manager knows, so why bother escalating? We do not have a way of turning down a resource or giving feedback to the manager or the individual. The functional manager has no idea how this person really functions. So you succeed by finding a horde of people you like and trying to synchronize the ending of one project with the beginning of your next so you can hang onto them." – *Project Leader, Manufacturing Firm*

Application

THE ROUNDUP CASE

The complete undoing of Roundup was not guaranteed by the existence of these problems but by how they were handled. Most were never fully discussed with the people who needed to hear them. Others were raised, but then muffled by the very executives who should have listened and taken action.

KEY FINDINGS

The crucial issues being addressed are so common that many leaders have stopped seeing them.

CONCLUSION

Although *Silence Fails* focuses on important findings that can predict and explain failure, the most important implication of the study is that potential leaders have to influence success.

While the data show that fewer than one in five project leaders effectively engage in the crucial conversations needed to solve these five problems, the good news is that one in five does. Much can be learned by studying their successes and whether these unique individuals produce different results than their stymied peers.

Research on this contingent of crucial conversations successes demonstrates that project leaders who speak up effectively do affect project success, and that others can be taught to speak up more skillfully and achieve similar results. Not only is it possible to rapidly and measurably change patterns of silence in an organization to patterns of open dialogue and accountability, it's also possible to sustain these results.*

Following are a few of the best practices project leaders and leaders can begin using immediately to address the issues raised in *Silence Fails*. Use these best practices to face the issues and begin building organizational competence at resolving them.

Getting People to Speak Up Well

As stated earlier in this report, in each of the five key problem areas, the study found a clear difference between speaking up and speaking up well. Across the problem areas, about half of leaders make some attempt to speak up. But most are ineffective. Some speak up but they water down their concerns, so the issues are never fully aired. Some speak up but do so in a way that provokes defensiveness from others. And a handful—are able to share their full concerns by the end of the conversation, feel their views are understood and respected.

The study also shows that while the skill of the initiator is a key ingredient in ensuring these crucial conversations are held well, the receptiveness of the other party is similarly important. Project leaders who are adept at holding crucial conversations are far more effective in dealing with prickly and defensive executives. But their skill don't guarantee that their concerns will be addressed. Unless and until leaders take extraordinary measures to ensure their environment is conducive to holding crucial conversations, a significant number of issues will remain unaddressed, invisible, and fatal.

1. Develop a Business Case for Change

The crucial issues being addressed are so common that many leaders have stopped seeing them. Rapid change can be made, but not if these crucial conversations are seen as "soft" issues—or as a "nice to do." Consider the following steps to develop your business case for change:

- Begin by making the problem visible. Track and publish data about project successes and failures.
- Distribute *Silence Fails* to generate discussion about the root cause of current underperformance.
- Engage senior leaders in a "listening campaign" where they lead structured focus groups to validate whether these crucial problems affect current results.

The goal of this process of published results, engaging discussion, and focused listening is to help senior leaders connect the behaviors described here with results they care intensely about. Once they can articulate this connection, they will see the business need for changing these vital behaviors.

2. Measure Behaviors

Leaders who are serious about improving how their people address these conversations are the first to regularly measure current behavior.

- Conduct a *Silence Fails Assessment* (find one at www.silencefails.com) to determine how well your project is performing in relation to the five crucial conversations in *Silence Fails.*
- Discuss the results of the *Silence Fails Assessment* with the right people. Whether it's the senior management team or another team of cross-functional leaders, this will help draw attention to the existence of the crucial issues and whether they are being adequately discussed and addressed.
- Use an organizational survey to assess executive competence within your orgainzation.

KEY FINDINGS

When a skillful leader creates even a moderately safe environment, the likelihood of project success increases 50 to 70 percent.

3. Invest in Skills

Most project leaders and team members lack the confidence to address these politically sensitive issues because they don't know how to lead this risky discussion. Leaders who train their people to deal with these specific crucial conversations see substantial improvement in whether and how the issues get resolved.

- Publish a bold improvement goal. Get senior leadership committed to specific improvements over baseline measures.
- Follow up with training, discussion, assessments, and other tools to ensure that transference is occurring. Monitor the impact of this training in your organization.
- Map training outcomes to key performance indicators. Determine the behaviors that are connected to the results you want to improve and track both the behavioral change and that bottom-line result you're after.

4. Hold Senior Management Accountable

Investing in project leaders' competence at holding crucial conversations is necessary but insufficient. Effective organizations judge leaders and hold them accountable for improving the measures on this survey within their areas of responsibility. The best even tie it to senior leadership bonuses and performance evaluations. In addition:

- Make leaders the teachers. People will change their behavior more rapidly if leaders deliver the training than if staff trainers or outside consultants do so. When leaders teach, the speed of change can be two or three times greater than when those who aren't as credible and connected in the organization lead instruction.*
- Hold everyone accountable for creating a safe environment for these crucial conversations.

When senior leaders know that they are accountable for how well they lead at changing people's behavior, they get much more interested.

5. Make Heroes of Early Adopters

The key to getting a hundred people to speak up is to publicly reward the first one who does. Be sure to send a clear and public message that these conversations aren't just important, they're *crucial*—and those who raise them are highly valued. Tips include:

- Highlight and reward people who take a risk and raise these crucial conversations.
- In a public forum, praise individuals who surface sensitive project risks or skillfully challenge leaders.

If you want to change the values of an organization away from silence and toward candid dialogue, make heroes of those who take a chance with these new behaviors.

For the resources recommended here and additional tools, please visit www.silencefails.com.

Section Notes

* Based on over 25 years of research, VitalSmarts continue to conduct field studies that demonstrate significant, sustainable results brought about through organizational change initiatives. These efforts focus on behavior modification as opposed to system design and strategy development. See www.vitalsmarts.com for case studies and additional information.

About the Sponsors

VitalSmarts

An innovator in corporate training and organizational performance, VitalSmarts helps teams and organizations achieve the results they care about most. With award-winning training products based on more than 25 years of ongoing research, VitalSmarts has helped more than 300 of the Fortune 500 realize significant results using a proven method for driving rapid, sustainable, and measurable change in behaviors. VitalSmarts has been ranked twice by Inc. magazine as one of the fastest-growing companies in America and has trained more than 500,000 people worldwide. www.vitalsmarts.com

The Concours Group

The Concours Group is a new breed of professional services firm, supporting senior executives through the blend of leading-edge intellectual capital and pragmatic business applications. The firm works with more than 300 of the Global 1,000 firms, helping leaders turn human and technological potential into business value. The Concours Institute is the research and education arm of The Concours Group. Concours research discovers and develops future best practices in business, technology, and human capital; its education articulates them and motivates their adoption; and its innovative senior-person Advisory Services enables clients to implement them quickly and achieve business results. www.concoursgroup.com

ABOUT THE STUDY

Beginning in late 2005, VitalSmarts and The Concours Group partnered to dive below the formal systems to identify what was missing. With worldwide participants, the study *Silence Fails* yielded substantial findings. The study was broken into two phases: exploratory and observational.

EXPLORATORY PHASE

Researchers conducted interviews and gathered statements from individuals in leadership positions, as well as individual contributors involved in project management. From there, the research team analyzed emerging themes and patterns to produce a set of issues that were most prevalent and costly. Results were retested with focus groups. Finally, a preliminary test of the survey was conducted to confirm that our factor analysis was consistent with a survey-based approach.

OBSERVATIONAL PHASE

The observational phase of our study, including a full survey, included the following:

- Researchers spent more than 150 hours observing various phases of project chartering, approval, and execution. In addition, interviews, focus groups, and surveys were administered to more than 1,000 senior executives, project sponsors, project leaders, and project participants.
- More than 1,000 executives and project management professionals across 40 companies were involved. Participating organizations came from a wide variety of industries, including pharmaceuticals, a fast food conglomerate, an international construction company, airlines, financial services, government agencies, and consumer products. While most of the 40 companies were Fortune 500 multinational organizations, about 10 percent were smaller, regional firms.
- The analysis encompassed more than 2,200 projects, ranging from $10,000 IT projects to billion-dollar organizational restructuring efforts.
- Study participants ranged in experience from veteran, university-trained and certified project leaders to newly minted managers with little formal training in project management. Some organizations had sophisticated project governance, management processes, and policies, while others had far less developed approaches.

APPENDIX

Aggregated study findings are outlined below.

		Fact-Free Planning	AWOL Sponsors	Skirting	Project Chicken	Team Failures
Extent of Problem	Project managers with this problem	85%	65%	83%	55%	80%
	% of their projects affected	39%	33%	33%	31%	30%
	Those reporting it's hard to confront/resolve.	91%	88%	79%	61%	76%
	Those who speak up at all	61%	47%	51%	42%	49%
	Those who speak up skillfully	14%	11%	13%	13%	14%
	Those able to solve the problem	19%	17%	16%	26%	24%
Cost of Not Speaking Up	The problem continues for the life of the project or beyond	88%	75%	81%	76%	80%
	Budgets are exceeded	82%	75%	78%	78%	73%
	Deadlines are missed	82%	85%	87%	86%	82%
	Functionality specs or quality suffers	82%	74%	80%	74%	77%
	Team morale is damaged	73%	69%	66%	54%	69%
	A long list of problems remain to be resolved after the project ended	43%	31%	28%	23%	24%
Impact of Speaking Up	Reduced cost overruns	29%	33%	35%	55%	64%
	Reduced schedule delays	17%	33%	28%	69%	60%
	Improved quality and functionality	69%	50%	44%	56%	64%
	Were less likely to take inappropriate shortcuts	47%	66%	64%	66%	66%
	Were less likely to have a long list of problems to resolve after the project ended	58%	70%	71%	70%	70%

For more information about this study, contact VitalSmarts at 800.449.5989 or via e-mail at research@vitalsmarts.com

SILENCE FAILS ASSESSMENT

Will Your Project Succeed?

Consider a current project or a project that you've recently completed.

Project Name: ____________________ Project Leader: ____________________

Section 1: Fact-Free Planning	Yes	No
1. Are the timelines, budgets, materials, and personnel established for this initiative realistic—given the deliverables required? (If you answer "Yes" on this question, go to the next section. If you answer "No," answer questions 1a–1c below.)	☐	☐
1a. If the resources are not realistic, have you fully discussed all of your concerns with the person(s) who could change them to make them realistic (e.g., project sponsors, resource managers, etc.)?	☐	☐
1b. Are you confident that the person(s) who can make the changes will take the actions required to adjust the resources, deliverables, or other elements of scope so that the initiative can succeed?	☐	☐
1c. If the person(s) who can make the changes to the resources or the deliverable or other elements of scope do so, will it be enough to make the project a success?	☐	☐

Score for Fact-Free Planning (add up the checks under "No"): ________

Section 2: AWOL Sponsors	Yes	No
2. Is your sponsor providing the kind of active involvement and support this initiative needs to be successful (e.g., convincing reluctant stakeholders, supporting you in policy meetings, working behind the scenes to keep the initiative on track)? (If you answer "Yes" on this question, go to the next section. If you answer "No," answer questions 2a–2c below.)	☐	☐
2a. If you are not getting the level of support you need from your sponsor, have you fully discussed all of your concerns with him/her?	☐	☐
2b. Are you confident that your sponsor will soon become as actively involved as you need him/her to be?	☐	☐
2c. Is your sponsor's active involvement going to be enough to make the project a success?	☐	☐

Score for AWOL Sponsorship (add up the checks under "No"): ________

Section 3: Skirting	Yes	No
3. Are the right stakeholders supporting the priority setting process with this initiative? (e.g, not exerting inappropriate influence; bypassing channels to change the scope, schedule, or resources; pulling rank to get their way; or cutting deals on the side in a way that threatens the initiative)? (If you answer "Yes" on this question, go to the next section. If you answer "No," answer questions 3a–3c below.)	☐	☐
3a. If a powerful person is exerting inappropriate influence on your project, have you discussed your concerns with him/her?	☐	☐
3b. Are you confident that this person will stop exerting his/her influence in inappropriate ways—e.g., bypassing channels, pulling rank, cutting deals, etc?	☐	☐
3c. If this person stops exerting his/her influence in inappropriate ways, will that be enough to make the project a success?	☐	☐

Score for Skirting (add up the checks under "No"): ________

Section 4: Project Chicken

	Yes	No
4. Are people giving you honest, accurate and timely information regarding the status of your initiative, allowing you to get a clear view of where it stands? (If you answer "Yes" on this question, go to the next section. If you answer "No," answer questions 4a–4c below.)	☐	☐
4a. If a person is failing to get you honest, accurate, or timely information on the status of the project, have you discussed your concerns with him/her?	☐	☐
4b. Are you confident that this person will start getting you honest, accurate, and timely information on the status of the project?	☐	☐
4c. If this person starts getting you honest, accurate, and timely information, will that be enough to make the project a success?	☐	☐

Score for Project Chicken (add up the checks under "No"): ________

Section 5: Team Failures

	Yes	No
5. Are team members performing up to your expectations and keeping commitments, and are they sufficiently competent to see the initiative through to the end? (If you answer "Yes" on this question, follow the instructions below to score your assessment. If you answer "No," answer questions 5a–5c below.)	☐	☐
5a. If a person is not performing up to your expectations (i.e., he/she does not have the right competencies, talents, time, etc.) or if a resource manager is not assigning qualified candidates, have you discussed your concerns with him/her?	☐	☐
5b. Are you confident that this person will begin performing up to your expectations?	☐	☐
5c. If this person begins to perform up to your expectations, will that be enough to make the project a success?	☐	☐

Score for Team Failures (add up the checks under "No"): ________

Section Score:

The section with the highest number of no responses is your weakest section.

Overall Score (add up the total score from each section):

0: Your initiative has a good chance of success.

1-4: Your initiative has a few issues, but if you deal with the problem areas, there is still a good chance that your initiative will succeed. Unfortunately, If you fail to address even one of these crucial conversations, there is a 61 percent chance that your project will go over in budget, schedule, and scope.

5-9: Your initiative is in jeopardy. You have one or more big problems and possibly several smaller problems that you need to address to get your project back on track.

10-14: You have three to four major conversations that you need to hold and hold well.

15-20: Speak up! Your initiative is doomed!

To find out how to get better, visit www.silencefails.com

AUTHOR INDEX

American Institute of Chemical Engineers
3 Park Avenue
New York, New York 10016-5991

ISBN: 978-1-60560-003-1